In his paintings, American artist Paul Jenkins has long explored abstract images having an affinity to those to be found in the universe. He sees parallels between the space in a canvas and space beyond this Earth. Both can be entered freely by the mind, and neither is open to real conquest. Jenkins believes that the exploration of the universe, like the expression of art, can "help us become more of what we are."

The paintings on the cover and divider pages are reproduced with the permission of the artist.

Cover: *Phenomena Waves Without Wind*, 1977
© Paul Jenkins

Exploring the Living Universe
A Strategy for Space Life Sciences

A Report of the NASA Life Sciences Strategic Planning Study Committee

June 1988
Washington, DC

NASA Life Sciences Strategic Planning Study Committee

Chairperson
  Frederick C. Robbins, M.D., *Case Western Reserve University*

Executive Secretary
  James H. Bredt, Ph.D., *National Aeronautics and Space Administration*

Alternate Executive Secretary
  Maurice Averner, Ph.D., *National Aeronautics and Space Administration*

Members
  Ivan L. Bennett, M.D., *New York University Medical Center*
  Gerald P. Carr, P.E., D.Sci., *CAMUS, Inc.*
  Sherwood Chang, Ph.D., *National Aeronautics and Space Administration*
  Michael Collins, *Michael Collins Associates*
  William DeCampli, M.D., Ph.D., *Stanford University Medical Center*
  Peter B. Dews, M.D., *Harvard Medical School*
  Arthur W. Galston, Ph.D., *Yale University*
  Bernadine Healy, M.D., *Cleveland Clinic Foundation*
  Carolyn L. Huntoon, Ph.D., *National Aeronautics and Space Administration*
  Thomas E. Malone, Ph.D., *University of Maryland at Baltimore*
  Francis D. Moore, M.D., *Harvard Medical School*
  Robert H. Moser, M.D., *The NutraSweet Company*
  Jay P. Sanford, M.D., *Uniformed Services University of the Health Sciences*
  William C. Schneider, D.Sci., *Computer Sciences Corporation*
  J. William Schopf, Ph.D., *University of California at Los Angeles*
  Peter M. Vitousek, Ph.D., *Stanford University*

Staff Associates
  Keith L. Cowing, *Universities Space Research Association*
  Ross Hinkle, Ph.D., *Bionetics Corporation*
  Mitchell K. Hobish, Ph.D., *MKH and Associates, Inc.*
  Lauren Leveton, Ph.D., *Lockheed Engineering and Management Services Company*
  Barry J. Linder, M.D., *Washington University Medical Center*
  Warren Lockette, M.D., *University of Michigan, Henry Ford Hospital*
  Carole O'Toole, *Science Applications International Corporation*
  Mark H. Phillips, Ph.D., *Cowell Hospital, University of California*
  Beryl A. Radin, Ph.D., *University of Southern California, Washington Public Affairs Center*
  Mark L. Schlam, *Consultant*
  Mathew R. Schwaller, Ph.D., *Science Applications International Corporation*

Dedication

James H. Bredt, Ph.D.
Executive Secretary of the Life Sciences
Strategic Planning Study Committee
1986–1988

The NASA Life Sciences Strategic Planning Study Committee dedicates this publication to the memory of our friend and colleague James Bredt, who died on April 25, 1988. As Executive Secretary of the Committee, Jim made significant contributions to its deliberations. Even in the last stage of a debilitating disease, he continued to participate in Committee activities. His courage, commitment, and professionalism will be long remembered.

CASE WESTERN RESERVE UNIVERSITY · CLEVELAND, OHIO 44106

May 27, 1988

Mr. Daniel J. Fink
Chairman, NASA Advisory Council
National Aeronautics and Space
  Administration (NASA)
600 Independence Avenue, SW
Washington, DC 20546

Dear Dan:

On behalf of the NASA Advisory Council's Life Sciences Strategic Planning Study Committee, I am pleased to forward herewith our final report. The Committee began its work in September 1986. It was charged with reviewing the status of Life Sciences within NASA, examining its goals, and suggesting ways and means of attaining these goals.

The Committee's findings and recommendations are presented in abbreviated form in the summary. More detailed information, which provides support for the Committee's conclusions, is contained in a series of papers that comprise the body of the report. Background on the Committee and its members is presented in the appendix.

The Committee is firmly convinced and cannot emphasize enough that a stronger Life Sciences program is an imperative if the U.S. space policy is to construct a permanently manned Space Station and achieve its stated goal of expanding the human presence beyond Earth orbit into the solar system. The same considerations apply in regard to the other major goal of Life Sciences: to study biological processes and life in the universe. Developing a stronger program will require increasing the involvement of first-rate investigators at both universities and research institutions as well as at the present NASA research centers, such as Ames and Johnson. Some of our recommendations deal specifically with these issues. It is evident that much depends upon varied flight opportunities. From the point of view of learning what is necessary so that man can exist safely for extended periods in space, however, the availability of the Space Station becomes crucial. Not only is the Space Station critical, but the facilities need to be adequate as to size and equipment to achieve their purpose. The complexity of the issues and the multidisciplinary nature of the Life Sciences enterprise require that life scientists be intimately involved in most aspects of NASA's overall planning and design activities, whether they concern setting budget priorities, developing the Space Station, designing space suits, or programming unmanned probes to Mars.

Cleveland Study of the Elderly
Department of Epidemiology
  and Biostatistics
School of Medicine
Area Code: 216 Telephone 368-3760

Mr. Daniel J. Fink
May 27, 1988
Page 2

The Committee recognizes that its recommendations, if implemented, could increase more than twofold the expenditures on life sciences activities. However, if indeed one of our priorities is to place man in outer space under conditions that are safe and yet permit an adequate quality of life and work, we see no alternative to a considerable expansion of the program directed to this end. In our opinion, the recent experience of a Soviet cosmonaut who spent over 300 days in space only highlights the need for well controlled and designed experiments to elucidate further the physiological and psychological effects of prolonged existence in space and to devise and test means to counteract them.

Although there is a tendency to emphasize manned space flight, we would regard the Mission to Planet Earth or the Biospherics Program also to be of paramount importance in view of the extraordinary manmade threat that promises to seriously and perhaps permanently imperil the ecological balance on Earth. It is through such programs as Biospherics that we can define the problem and approach solutions.

The Exobiology Program and parts of the program of Gravitational Biology are directed at problems of great intrinsic scientific interest.

We would further wish to emphasize that, like much of the research conducted by NASA, life sciences research will contribute to scientific knowledge irrespective of its applicability to the specific needs of the space program.

Finally, I should like to comment upon the importance the Committee placed upon international cooperation. Because of the many other topics addressed in the report, this one may not command the reader's attention to the degree it deserves. The Committee feels that much mutual benefit can be derived from expanding true international cooperation despite the difficulties this may entail. Increased interaction with the U.S.S.R. could be particularly valuable in view of their more extensive experience with man in space for prolonged periods.

It has been gratifying to observe that, as we have been in the process of developing our report, certain changes have occurred in the Life Sciences program that have anticipated some of our recommendations. Thus, the program is stronger today than it was 18 months ago.

The Committee is appreciative of the opportunity to conduct this study. We wish to acknowledge the assistance and cooperation of the scientific community at large, those Federal agencies involved, and, especially, the NASA Headquarters and field centers, which gave so generously of their time and information. Furthermore, I personally wish to thank the Committee members and the Staff Associates for their extraordinary efforts in making this report possible.

Respectfully,

Frederick C. Robbins, M.D.
University Professor Emeritus,
Dean Emeritus, School of Medicine

National Aeronautics and
Space Administration

Washington, D.C.
20546

Office of the Administrator

June 3, 1988

Honorable James C. Fletcher
Administrator
National Aeronautics and Space Administration
Washington, DC 20546

Dear Jim:

I am pleased to forward with this letter the report of the NASA Advisory Council's Life Science Strategic Planning Study Committee. The report, "Exploring the Living Universe: A Strategy for Space Life Sciences," is the product of an intensive study by a group of renowned experts in various life science and other disciplines. It addresses and provides recommendations on goals, objectives, and priorities for the overall life science program; for the sub-programs of human space flight, gravitational biology, and planetary biological research; for flight programs; and for program administration. When presented to the full Council at its meeting on May 25, 1988, it was enthusiastically endorsed and approved for transmittal to you.

One principal recommendation of the report is for NASA to expand its program of ground- and space-based research contributing to resolving questions about physiological deconditioning, radiation exposure, potential psychological difficulties, and life support requirements that may limit stay times for personnel on the Space Station and complicate missions of more extended duration. Other key recommendations call for strengthening programs of biological systems research in: controlled ecological life support systems for humans in space, Earth systems central to understanding the effects on the Earth's environment of both natural and human activities, and exobiology. The Council has long supported strengthening space life science programs and our concerns voiced in prior reports to NASA were in large measure responsible for commissioning this study.

This report joins those of the Solar System Exploration Committee and the Earth Systems Science Committee as keystones for planning the respective programs for some years to come. Fred Robbins and his committee members, associates, and staff have earned NASA's and the Council's commendations and thanks for a job well done.

Sincerely,

Daniel J. Fink, Chairman
NASA Advisory Council

Enclosure

Foreword

NASA is contemplating a future in space that would include permanent human colonies on the Moon and Mars, as well as automated probes into the solar system and studies from space of Earth systems. Before such efforts can be attempted, the Agency must resolve life sciences issues central to the success of the U.S. civilian space program.

To identify these issues, the NASA Advisory Council (NAC) authorized the establishment of the Life Sciences Strategic Planning Study Committee (LSSPSC) in the spring of 1986. Organized in the following summer under the chairmanship of Dr. Frederick C. Robbins, the LSSPSC was charged with developing a comprehensive view of space program issues related to the life sciences, recommending goals for NASA's life sciences efforts, and devising feasible scientific and technical strategies to achieve these goals.

The LSSPSC presents the results of its research, including findings, recommendations, and a strategy for life sciences, in this report. The study is the third in a series commissioned by the NAC on major parts of the space program. The earlier publications were *Planetary Exploration Through Year 2000: A Core Program* (NASA, 1983), developed by the Solar System Exploration Committee, and *Earth System Science: A Closer View* (NASA, 1987), drafted by the Earth System Sciences Committee.

In presenting a global view of life sciences at NASA, the LSSPSC report focuses on programs and issues that cut across disciplinary and organizational lines. The Life Sciences Division, the organizational center for life sciences activities at the Agency, is multidisciplinary in approach, incorporating activities that extend from basic science to clinical applications. Programmatic research concentrates on needs fundamental to human space flight, on the intricate workings of Earth as a biosphere, and on the possibilities of life past, present, and future in the universe.

The challenge for the Life Sciences Division lies in its multidisciplinary approach, which necessitates the ongoing integration of contributions from various scientific areas and sponsoring organizations. The value of its programs to NASA comes from this same approach, designed to meet certain of the Agency's diverse requirements. Life sciences research is basic to establishing the capabilities for safe and productive, long-term human activity in space, to developing human communities on other planets, to exploring the origin, evolution, and distribution of life in the universe, and to reestablishing U.S. leadership in civilian space endeavors.

Contents

Foreword ix

Summary 1

1. Overview 17
   Space Medicine and Biology 18
   Biological Systems Research 19
   Flight Programs 21
   Future Course 22

2. Findings and Recommendations 25
   Overarching Recommendations and Strategies 25
   Human Space Flight 28
   Gravitational Biology 31
   Planetary Biosciences Research 33
   Flight Programs 35
   Program Administration 37

3. Life Sciences in the Space Program 39
   Biomedical Research 40
   Radiation 53
   Crew Factors 67
   Systems Engineering 79
   Operational Medicine 91
   Gravitational Biology 101
   Controlled Ecological Life Support Systems 112
   Biospherics Research 124
   Exobiology 132
   Flight Programs 154
   Program Administration 171
   Applications 185

Appendix 191
   Background on the Committee 191
   Glossary 201
   Selected Bibliography 203
   Photograph Credits 212

Index 213

List of Abbreviations and Acronyms .... inside back cover

Summary

Lunar Moth, 1958
© Paul Jenkins

Summary

Visionaries have long speculated over a future in which humans understand the scientific truth about their origins, control their environment on Earth, and live successfully outside of that environment. Their speculations, however, have frequently overlooked some fundamental facts: that the universe is complex and mostly inhospitable and that life as we know it evolved in the protective shelter of an atmosphere and a constant gravitational force.

The knowledge obtained by space life sciences will play a pivotal role as humankind reaches out to explore the solar system. To conduct the types of space missions contemplated by the National Aeronautics and Space Administration (NASA), information is needed concerning the existence of life beyond the Earth, the potential interactions between planets and living organisms, and the possibilities for humans to inhabit space safely and productively.

Our experience in space thus far has given us a glimpse of the potential problems and rewards facing humans on future missions, particularly those of long duration. Within the United States space program, NASA life sciences are responsible for acquiring knowledge that will contribute to the human exploration of space. Programs in the involved disciplines are an integral part of NASA's current and future missions, from near Earth orbit, to human missions to the Moon and Mars. To realize their objectives, they require the development and operation of diverse ground and flight facilities and close coordination with numerous scientific and governmental organizations in the United States and abroad.

Study Committee Charge

Given the need for a vigorous and forward-looking program in the space life sciences, Dr. James Fletcher, the NASA Administrator, charged the NASA Advisory Council (NAC) with developing a strategic plan that will prepare the Agency for the approaching era of space exploration. To accomplish this task, the NAC organized the Life Sciences Strategic Planning Study Committee (LSSPSC) under the leadership of Frederick C. Robbins, M.D.

The LSSPSC pursued its work within a context shaped by the reports of recent task groups: *Leadership and America's Future in Space* (NASA, 1987), *A Strategy for Space Biology and Medical Science for the 1980s and 1990s* (National Academy of Sciences, 1987), and, among others, *Pioneering the Space Frontier* (National Commission on Space, 1986). Many of the issues discussed in these publications were relevant to the objectives of the LSSPSC, which cited the volumes on several occasions. The Committee, however, considered these matters independently and made efforts to avoid duplication of activity. The findings and recommendations in the LSSPSC report are consistent with those given in the other task group publications, particularly in *A Strategy for Space Biology and Medical Science for the 1980s and 1990s*, and with the National Space Policy, issued by President Ronald Reagan in February 1988. **To reassert U.S. leadership in space research and exploration, it is vital that life sciences be an integral part of the Nation's space program.**

The NASA Space Life Sciences Program

Gravitational Biology

- Understanding the role of gravity in the development and evolution of life

Biomedical Research

- Characterizing and removing the primary physiological and psychological obstacles to extended human space flight

Environmental Factors

- Defining the space environment and habitat in which humans must function safely and productively, including air and water quality and the biological effects of radiation fields

Operational Medicine

- Developing medical and life support systems to enable human expansion beyond the Earth and into the solar system

Biospherics Research

- Developing methods to measure and predict changes on Earth on a global scale and the biological consequences of these changes

Physicochemical and Bioregenerative Life Support Systems

- Assembling the knowledge base needed to design, construct, and operate life support systems and extravehicular suits in space that are independent of major resupply

Exobiology

- Exploring the origin, evolution, and distribution of life in the universe

Flight Programs

- Conducting experiments in space, including the development of facilities and hardware for space flight, mission planning integration, and flight plan implementation

Overarching Recommendations and Strategies

In its deliberations, the Committee recognized that the resolution of certain key factors was pivotal to the foundation of vigorous life sciences programs. It stresses the importance of these factors by incorporating them in a set of overarching recommendations. The Committee recommends that the Agency should:

- **Maintain and expand the Nation's life sciences research facilities located at NASA's field centers, universities, and industrial centers by:**
  - Establishing a mechanism for attracting promising young scientists to work on NASA projects and developing additional training programs at major universities and NASA installations
  - Establishing a program of NASA-supported professorships in space life sciences at selected universities
  - Encouraging industries to develop capabilities in space life sciences through technology research and development.

- **Assure timely and sustained access to space flight, thereby facilitating the conduct of critical life sciences experiments. This should be accomplished through:**
  - Accumulating state-of-the-art instrumentation
  - Flying an augmented series of Spacelab missions
  - Using a series of autonomous bioplatforms to study radiation and variable-gravity effects
  - Dedicating suitable facilities on the Phase 1 Space Station for life sciences research
  - Conducting a major augmentation of life sciences capabilities during the early Post-Phase 1 Space Station.

- **Synergize the presently independent research activities of national and international organizations through the development of cooperative programs in the life sciences at NASA and university laboratories.**

- **Complete and consolidate the unique national data base consisting of basic life sciences information and the results of biomedical studies of astronauts conducted on a longitudinal basis. This data base should be expanded to incorporate information obtained by other spacefaring nations and be available to all participating partners.**

To achieve these recommendations, NASA should initiate work immediately, in the 1989 fiscal year, on the following set of strategic milestones:

1989-1991

- **Strengthen the planning process of the Life Sciences Division by assuring its timely integration into the Agency's overall strategic planning process.**

- Augment life sciences research programs to establish the base of scientific knowledge required by planners and engineers to conduct missions relevant to Agency goals.

- Provide adequate funding to develop new state-of-the-art flight hardware for upcoming manned and unmanned life sciences missions in space.

- Initiate advanced technology development in the areas of minimally invasive biomedical instrumentation, biological remote sensing, exobiological flight instrumentation, and microwave signal processing.

- Increase the frequency of life sciences data acquisition on the Space Shuttle and international missions.

- Conduct a study to determine the requirements for extravehicular activity (EVA) for the next 20 years, to delineate innovative options, and to identify needed technologies.

1989-1994

- Operate reusable biosatellites to obtain environmental, radiation, and artificial variable-gravity data on plants and animals.

- Achieve ground-based validation of major physiological and psychological countermeasures for long-duration missions.

- Conduct ground-based research on bioregenerative life support systems to achieve 90-percent closure.

- Initiate the Microwave Observing Project of the Search for Extraterrestrial Intelligence (SETI) Program.

1989-2004

- Establish a combined national and international life sciences research facility on the Space Station. This facility must support basic research on plants, animals, and humans necessary to develop an understanding of the fundamental biological processes affected by gravitational forces.

- Develop an advanced biomedical research facility in space to investigate and verify technologies and medical support necessary to enable the planning and implementation of human exploration of the solar system.

- Develop and test in space a fully operational bioregenerative life support system(s) for future use in solar system exploration.

- Conduct cooperative missions with other national and international organizations to study the behavior of the biosphere and the origin, evolution, and distribution of life on Earth and in space.

The strategic milestones emphasize the importance of international cooperation in space life sciences research and missions. The LSSPSC believes that considerable mutual benefit can be derived from expanding such efforts. Increased interaction with the U.S.S.R. could be particularly valuable because of their more extensive experience with humans in space for prolonged periods.

The LSSPSC also discussed the need to quantify resources, including the funding, personnel, and facilities required for implementation of the strategic milestones. It determined that this activity was critical but could not be satisfactorily accomplished in the time available to the Committee. The LSSPSC accordingly recommends that this effort be initiated immediately after publication of the report through techniques and resources readily available to NASA and that the results be communicated as available to the NASA Advisory Council.

Implementation of the strategic plan requires the careful scheduling of activities relevant to the two major program thrusts:

- The assurance of the health, safety, and productivity of humans in space
- The acquisition of fundamental scientific knowledge concerning space life sciences.

These emphases are equally important, the first being an Agency goal and the second being a part of the strategic plan developed by the NASA Office of Space Science and Applications (OSSA). Efforts associated with assuring the health, safety, and productivity of humans in space should be paced so as to provide the Agency with information vital in planning and conducting extended manned missions. While much can be done using ground research and short-duration flights, the key lies in the availability of appropriate life sciences facilities on the Space Station. Scheduling pertinent to the basic scientific programs should be consistent with the OSSA overall long-range strategic plan.

Committee Deliberations

The LSSPSC organized into 13 Study Groups to evaluate NASA life sciences activities. The Study Groups surveyed scientific literature, interviewed NASA researchers and administrators, and deliberated with international authorities from Europe, Japan, and the Soviet Union. The groups summarized the results of their research in papers that provided the basis for the Committee's findings and recommendations.

Based on the Study Group evaluations and research papers, the Committee developed approximately 30 detailed recommendations in addition to the four overarching recommendations. The detailed recommendations appear below under the headings of Human Space Flight, Gravitational Biology, Planetary Biosciences Research, Flight Programs, and Program Administration.

Specific Recommendations and Findings
Human Space Flight

Four challenges potentially limit the duration of human space flight: physiological deconditioning, the biological effects of exposure to ionizing radiation, possible psychological difficulties on the part of the space crew, and environmental requirements, including the need of life support on lengthy space journeys. The disciplines of Biomedical Research and Operational Medicine focus on the health and safety of human space crews.

Biomedical Research concentrates on physiological deconditioning, which becomes a greater concern the longer the space mission. Ground and space research have identified unresolved scientific issues relevant to the following areas: cardiovascular physiology, specifically, a more complete characterization of cardiovascular deconditioning; neurophysiology and behavioral physiology, particularly space adaptation syndrome (space motion sickness); bone, endocrine, and muscle physiology.

> ### *Soviet Space Accomplishments*
>
> The recent return of Soviet cosmonaut Yuri Romanenko to Earth after 326 days in space has excited great interest, as evidenced by reports in the world press. His return suggests that humans can exist for considerable periods in space and successfully readapt to conditions on Earth.
>
> Caution must be exercised, however, in drawing optimistic conclusions from a single case, particularly when the subject was unusually experienced in space missions and had been selected according to particular physiological and psychological attributes.
>
> Furthermore, the assertion that regular exercise played a role in preserving his well-being has yet to be proved. It should be noted, in addition, that his exercise program consumed 4 hours each day.
>
> Thus, while Romanenko's experience is encouraging, it only makes more imperative that we pursue as soon as possible the necessary studies in space to define better the physiological changes over time so that countermeasures can be rationally devised.

Radiation poses significant challenges for long-duration missions, such as the 1 to 3 years required for a round trip to Mars. While considerable information is available about radiation beyond the protection of Earth's magnetic field, substantive questions remain concerning the biological effects of exposure to galactic cosmic radiation and solar particle events and the shielding required to protect astronauts, as well as exposure-measuring instrumentation. Although critical unresolved issues remain, NASA does not have a focused program of radiation effects studies.

The success of extended missions will depend substantially on the psychological interactions among the space crew and between the space and ground crews. Information is not available on morale and productivity among small, isolated groups living in microgravity for lengthy periods. The most pressing issues for extended human missions, which will offer only limited possibilities for emergency rescue and return to Earth, involve crew/environment interactions, interpersonal interactions, human/machine interface, crew selection, command and control structure, and crew motivation.

Environmental factors and life support requirements directly relate to both the physiological and psychological well-being of the space crew. The primary concerns in this area include identifying requirements for a regenerative food, air, and water system, developing an environmental monitoring system capable of detecting all possible sources and types of contamination, determining the most workable systems to support EVA operations, and analyzing habitability requirements for extended missions.

The development of a bioregenerating life support system is especially challenging. NASA's Controlled Ecological Life Support Systems (CELSS) Program focuses on combining biological and physicochemical processes to provide food, air, and water by recycling materials inside the spacecraft. Ground-based research indicates that such a system is feasible. The behavior of plants in space, however, is not well understood.

Operational Medicine considers the health care of astronauts, particularly during long-duration missions. The most important operational issues include the development of requirements for the Health Maintenance Facility (HMF), definition of medical requirements for a Crew Emergency Return Vehicle (CERV), development of a data base for astronaut health records, and establishment of training programs for inflight medical specialists.

Recommendations: In addressing the ground- and space-based research needed to resolve the outstanding issues pertinent to human space missions of extended duration, NASA should:

- Immediately expand its program of ground-based research to resolve the outstanding questions about physiological deconditioning, radiation exposure, potential psychological difficulties, and life support requirements that may limit stay times for personnel on the Space Station and more extended missions.

- Plan an orderly, phased introduction of advanced life support and EVA technology into future manned space systems.

- Design and build a suite of variable-gravity facilities for life sciences research.

- In allocating payload and support resources for the Space Station, give first priority to life sciences research that will make human missions of extended duration possible.

- Take a number of steps, including the following, to ensure crew health and safety on the Space Station and missions of longer duration: include a physician among the crew, develop a Crew Emergency Return Vehicle to allow transport of crew members to Earth in urgent situations, and develop the capabilities of the Health Maintenance Facility for use on a possible human mission to Mars.

Gravitational Biology

Gravitational Biology studies the scope and operating mechanisms of one of the strongest factors influencing life on Earth: gravity. It addresses fundamental questions concerning how living organisms perceive gravity, how gravity is involved in determining developmental and physiological status, and how gravity has affected evolutionary history.

While these questions are motivated primarily by scientific interest, such research can help determine if life can function effectively for extended periods in weightlessness or reduced gravity, as on the Moon or Mars, or if artificial gravity is required. Space-based research, which requires variable-force centrifuge facilities, provides unparalleled opportunities to expose organisms to fractional gravity levels ranging from zero to 1 g, and thereby to investigate the effects of gravity on these organisms.

Ground-based vestibular sled experiments at NASA's Johnson Space Center test human response to rectilinear acceleration.

Recommendations: In understanding the role of gravity in the reproductive, developmental, and metabolic activities of all forms of life, NASA should:

- Increase the number, duration, and regularity of life sciences experiments flown in space.

- Provide adequate inflight research capabilities, including a suite of variable-force centrifuge facilities, on-orbit analytical equipment, and plant and animal vivaria capable of supporting successive generations subjected to varying, controlled gravity levels.

- Coordinate Gravitational Biology research with that conducted by interrelated science programs, such as CELSS and Space Biomedicine.

- Operate its intramural and extramural research programs in a manner that attracts and supports excellent new researchers, especially young scientists, into the relatively new field of Gravitational Biology, as well as into other areas of space life sciences.

Planetary Biosciences Research

The Biospherics Research Program studies the biological processes that have shaped the chemical history of Earth. Human activities, such as fossil fuel combustion, have markedly increased the concentrations of many atmospheric constituents, including greenhouse gases. A descriptive theory of the biosphere is required to understand the causes and consequences of these changes and to permit change measurement and prediction. Space capabilities are essential to this effort because they provide a global perspective.

The funding and logistical support needed to achieve biospherics goals transcends the resources of any single organization. Increased cooperation is, therefore, required among NASA organizations, Government agencies, and spacefaring nations.

Exobiology focuses on questions long pondered by humankind, such as, Are we alone in the universe? What led to the origin of life on Earth? Exobiologists believe that the early environments of Mars and Earth were similar and that samples from Mars could fill gaps in Earth's geological record. Any valid indication of life on Mars would support the hypothesis that life can originate wherever the physical and chemical environment is favorable. For these reasons, robotic probes followed by human missions to Mars will yield important scientific answers.

Recommendations: To understand the exobiology and biospherics issues relevant to the origin, evolution, and distribution of life in the universe, NASA should:

- Make the science requirements of biospherics and exobiology integral to plans for its Mission to Planet Earth and Exploration of the Solar System initiatives.

- Develop within those divisions having similar interests in planetary biology — the Life Sciences, Solar System Exploration, Earth Science and Applications, and Astrophysics Divisions — additional programs to promote maximum return from collaborative research.

- Include the Biospherics Research Program as a participant in the development and implementation of the Earth Observing System and other remote-sensing technologies.

- Initiate the Microwave Observing Project now, before radiofrequency interference makes it exceedingly difficult or impossible to conduct research from Earth.

- Pursue vigorous programs of ground-based research, remote observations, and instrument development for use on missions to assess evidence bearing on the possible origin of life on Mars and the nature of chemical evolution on other solar system bodies.

- Develop the technology of robotic round trip, sample selection and analysis, and sample return for exploration of the surface of Mars, asteroids, and comets. This effort should include precautions to avoid the spread of contamination within the solar system.

- Significantly enhance the ground- and space-based research capabilities and infrastructure (funding, personnel, and facilities) for planetary biology in order to maintain the Agency's leadership role in this area and to optimize the scientific return of future missions.

Flight Programs

Flight Programs includes the development of equipment, facilities, expertise, and flight opportunities needed to conduct life sciences research successfully in space. The hiatus in flight activity following the *Challenger* accident has been discouraging to life sciences researchers, many of whom have waited 10 years or more to fly their experiments. The current challenge is to assure that a sufficient number and variety of flight opportunities are available for life sciences research when the Shuttle resumes operations.

An additional challenge is to pursue a vigorous ground program that is closely integrated with and supportive of the flight program through significant ground preparations. These preparations include the design of equipment and the development of models that replicate space phenomena.

Astronaut Harrison Schmitt explores the huge lunar boulder during Apollo 17 *extravehicular activity at the Taurus-Littrow landing site.*

Recommendations: To facilitate the achievement of NASA and Life Sciences objectives, NASA should:

- Increase the flight opportunities for life sciences research by doing the following:
  - Dedicating a greater number of regularly scheduled Shuttle middeck lockers and commercially developed flight facilities
  - Increasing the flight rate of Spacelab and dedicating a larger percentage of Spacelab volume, time, and resources for life sciences experimentation
  - Dedicating clinical and biological research centers on the Phase 1 Space Station
  - Deploying an unmanned spacecraft that is reusable and can support a variety of flight experiments.
- Encourage students and non-NASA life scientists to participate in mission-related research but be careful not to encourage unrealistic expectations of flight opportunities.
- Develop a new generation of ground-based and of flight-certified instrumentation, including noninvasive monitoring techniques for biomedical applications, to support the research objectives of the Life Sciences program.

Program Administration

The administration of the Life Sciences program poses several difficult challenges. Because it encompasses basic science, applied science, operations, and engineering/technology activities, its management involves complex relationships that extend beyond disciplinary and organizational lines. Increased collaboration within NASA and stronger ties with universities are needed.

Although the Agency is committed to life sciences goals and objectives, the challenge of realizing these goals requires a major increase in Division resources. The provision of these needed resources at the proper time to permit required program growth is a critical issue. Increased budgets both in annual funding and for civil service personnel will enable the achievement of program objectives.

Recommendations: To strengthen the administration and organization of the life sciences, senior NASA management should:

- Support the continuation of Division efforts to establish a strong program by:
  - Strengthening the Division's role in Agency-wide planning
  - Facilitating access to flight opportunities
  - Indicating to the rest of the Agency that biomedical research relevant to the safe conduct of human space flight is essential to ongoing and future NASA initiatives.
- Include senior personnel from the Life Sciences Division as participants in all top-level planning of Agency flight programs.
- Increase substantially the resources for Life Sciences programs to assure implementation of the recommendations given in this report.
- Increase Agency efforts to expand the numbers of scientists at the Centers and Headquarters and institute new efforts to provide career development opportunities for existing staff.
- Support the Life Sciences Division in its efforts to establish formalized agreements and working groups with other agencies and organizations.
- Provide funds to expand and implement plans to establish Specialized Center of Research (SCOR) units within selected universities.
- Support the Life Sciences Division in generating and maintaining a data base through collaborative arrangements with NASA's Scientific and Technical Information Facility and the National Library of Medicine.

Abstracts of Topical Studies

Of the 13 LSSPSC Study Groups, 6 conducted indepth evaluations of NASA programs: Biomedical Research, Operational Medicine, Gravitational Biology, Biospherics Research, Exobiology, and Controlled Ecological Life Support Systems. The remaining seven groups investigated issues relevant to a number of programs and scientific disciplines: Radiation, Crew Factors, Systems Engineering, Flight Programs, Infrastructure, External Relations, and Applications. The Study Groups summarized their findings and recommendations in the papers given in section 3 of this report.

Summary

The first four papers discuss the four factors that potentially limit the duration of human space flight.

- "Biomedical Research" concentrates on physiological deconditioning, which becomes a greater concern the longer the space mission. The commentary addresses the unresolved scientific issues relevant to the following areas: cardiovascular physiology, specifically a more complete characterization of cardiovascular deconditioning; neurophysiology and behavioral physiology, particularly space adaptation syndrome (space motion sickness); bone, endocrine, and muscle physiology, involving total body calcium losses reportedly averaging 3 percent per month during space flight; and hematology, including the significant decreases in red cell mass reported in the Gemini, Apollo, Skylab, and *Soyuz* flight crews.

- The "Radiation" paper notes that NASA does not have a focused program of studies concerning radiation effects on space crews. Critical, unresolved issues remain in this field, however, particularly in connection with extended missions, such as a lunar colony, a manned round trip to Mars, and a Mars colony. Much more needs to be known about solar particle events, which could expose a space crew beyond the Earth's magnetic field to debilitating or lethal doses of radiation; the biological effects of high linear energy transfer (LET) radiation, including galactic cosmic rays; radiation-shielding interactions, which produce secondary radiation; and the most effective instrumentation for measuring real-time and cumulative radiation exposures, as well as changes in biological endpoints.

- The placement of the "Crew Factors" paper, immediately following discussion of biomedical and radiation concerns, emphasizes that human space flight involves not only physiological but also psychological survivability. The most pressing issues for extended manned missions, which will offer only limited possibilities for emergency rescue and return to Earth, involve crew/environment interactions, interpersonal interactions, human/machine integration, crew selection, command and control structure, and crew motivation. Answers to the related questions are beyond our direct experience, since the horizons are only beginning to open on long-duration human missions. For the rest of this century, ground-based research will be a practical mode for controlled experiments on group behavior.

- "Systems Engineering" discusses life support requirements, a factor directly related to both the physiological and psychological well-being of the space crew. The primary concerns in this field, which incorporates a broad range of disciplines, include identifying the requirements for a regenerative food, air, and water system that could support long-duration missions, developing an environmental monitoring system capable of detecting all possible sources and types of contamination, determining the most workable systems to support EVA operations, analyzing habitability requirements (such as the amount of space required per person) for extended missions, and identifying the requirements for a variable-gravity facility, such as a centrifuge.

The next five papers review the efforts associated with particular NASA programs. The first three explore issues pertinent to the effects of the space environment on fundamental biological processes and the factors potentially limiting human exploration of the solar system. The following two papers discuss issues related to the origin, evolution, distribution, and function of life on Earth and in the universe.

- "Operational Medicine" considers the health care of astronauts, particularly during long-duration missions. It proceeds from the understandings that no mission with humans in space can be risk free and that the goal of the Operational Medicine Program must be health risk reduction to a clearly defined level acceptable to the Agency. The most important issues in this area include the periodic review and revision of requirements for the Health Maintenance Facility, definition of medical requirements for a Crew Emergency Return Vehicle, development and maintenance of a data-base management system to incorporate inflight and ground-based medical records for astronauts, and the design of a training program for inflight medical specialists.

- "Gravitational Biology" explores the major issues in a field that has emerged with the advent of space flight. The discipline studies the scope and operating mechanisms of one of the most obvious and major environmental factors on this planet: gravity (g). Space-based research provides unparalleled opportunities to expose organisms to fractional gravity levels ranging from zero to 1 g and thereby to investigate the effects of gravity on these organisms. By so doing, it can help determine if humans, other animals, and plants can live and function effectively for extended periods in weightlessness or reduced gravity, as on the Moon or Mars, or if they require exposure to artificial gravity. Such research depends on the availability in space of a suite of variable-force centrifuge facilities.

- The "CELSS" paper, like the preceding discussion, notes the parallel emphases of the Gravitational Biology and the CELSS Programs, both of which conduct plant research and require access to space for key investigations. Both papers recommend collaborative efforts between the programs in areas of mutual interest. The long-term goal of CELSS is to create an integrated, self-sustaining system capable of providing food, potable water, and a breathable atmosphere for space crews on extended missions. Among the many issues requiring space research are the effects of weightlessness on plant growth, development, and reproduction. For extended human space missions to be possible early in the next century, the specific criteria for a CELSS need to be established well before the end of this century.

- "Biospherics Research" addresses the programmatic goals of the Biospherics Research Program, which are to develop methods to measure and predict changes to planet Earth on a global scale and the biological consequences of these changes. The funding and logistical support needed to meet these goals on a long-term basis transcend the resources of any single organization. The paper accordingly stresses the necessity of cooperation: between Biospherics Research in the Life Sciences Division and Terrestrial Ecosystems in the Earth

Science and Applications Division, the two NASA programs with primary responsibility for sponsoring global biogeochemical studies; among agencies that support biospherics research, including the National Oceanic and Atmospheric Administration and the National Science Foundation; and among international organizations, through such an effort as the International Geosphere-Biosphere Programme.

- "Exobiology" examines large questions that involve the origin, evolution, and distribution of life in the universe, that have sufficient drama and scope to elicit public attention and support, and that require for their resolution not only robotic probes of other planets but human exploration of the solar system, with the Moon and Mars as targets for early in the next century. The Exobiology Program is an interdisciplinary effort with interests complementary to Biospherics Research and other NASA programs in the Solar System Exploration and Astrophysics Divisions. By reviewing programmatic components, the paper identifies major scientific issues in gaining an understanding of the inter-relationship of life and the physical universe.

The final three Study Group papers — "Flight Programs," "Program Administration," and "Applications" — are not easily categorized, for their topics diverge considerably. The papers are alike, however, in that they all relate to the entire sweep of Life Sciences programs.

- The thesis of "Flight Programs" pertains to a primary emphasis in the LSSPSC report: the need for increased flight opportunities for life sciences. The hiatus in flight activity following *Challenger* has been discouraging to both established researchers and graduate students in life sciences, as indicated in the "Overview." Since flight opportunities will continue to be limited, even after resumption of Shuttle activity, the Life Sciences Division will have to conduct significant preparation on the ground, including designing and testing equipment and developing models that replicate space phenomena. At the same time, it needs to pursue a varied group of flight projects, including a recoverable, reusable biosatellite that can support flight experiments, as well as space in the Shuttle middeck lockers, Spacelab, and commercially available research facilities.

- "Program Administration," the one paper developed by two Study Groups (Infrastructure and External Relations), delineates the complexity involved in coordinating research that extends beyond disciplinary and organizational lines. Like several other Study Group papers, it emphasizes the need for additional collaboration between Life Sciences programs and other NASA offices, domestic agencies, and international organizations involved in the field. It also stresses the desirability of strengthening ties with universities, partly by establishing Specialized Center of Research units and NASA-supported professorships in space life sciences at selected institutions. At the same time, the paper recognizes that the Life Sciences Division does not have sufficient resources to meet its objectives, even through further cooperative efforts, and recommends a substantial increase in budget and provision for hiring additional civil servants.

- "Applications" concentrates not on the major goal of NASA, space exploration, but on an important secondary aim: applications research and technology transfer. Noting that NASA programs have generated over 30,000 documented spinoffs, the paper defines Federal and Agency policy supporting space applications. It then identifies the 16 Commercial Centers for the Development of Space and three university centers involved in life sciences applications, all established since 1985. While the Life Sciences Division does not specifically support technology transfer projects, commercial interfaces have been effectively implemented with life sciences personnel at the Centers. The paper suggests that a lead individual should be appointed at NASA Headquarters to receive questions on commercialization and refer them to the appropriate Centers.

The U.S. civil space program has reached a significant threshold. This country is contemplating a future in space that will involve increasingly complex missions that may include human bases on the Moon and Mars, as well as intensified satellite observation of Earth. NASA has recognized the critical role of the life sciences in facilitating and participating in such missions and must now commit the resources required to implement the strategy given in this report.

1. Overview

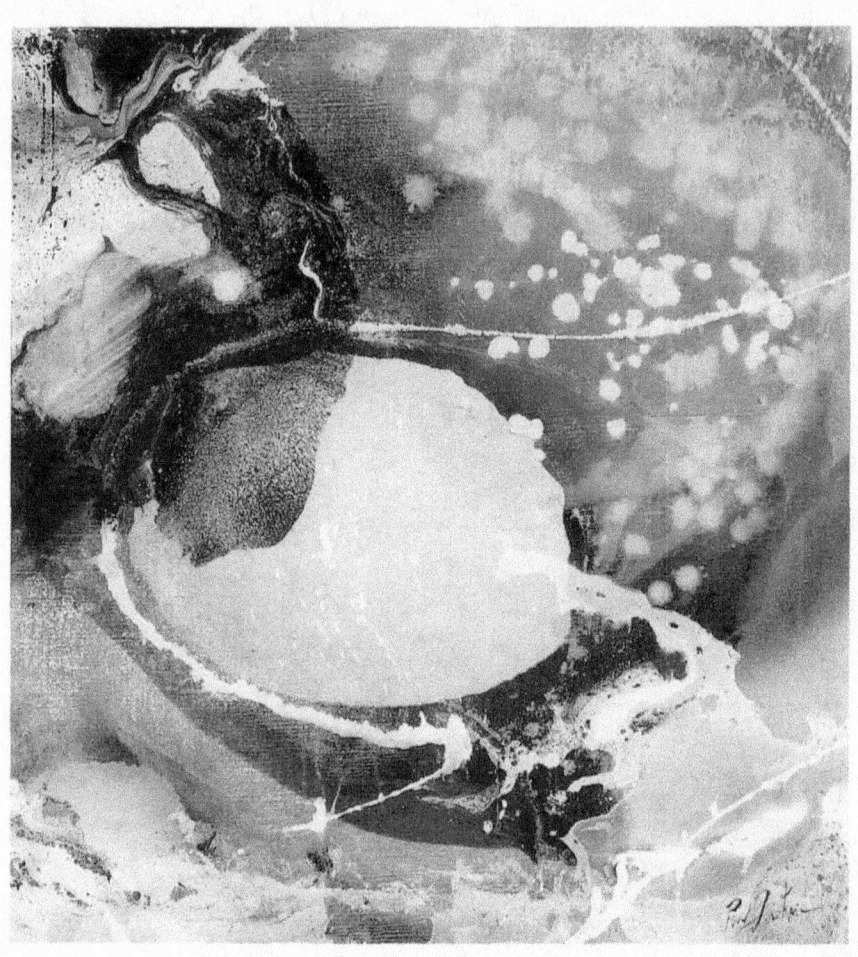

Uranus, 1956
© Paul Jenkins

1. Overview

NASA is engaged in a long-term quest for knowledge about life in the universe: its origin, evolution, and future, its distribution on Earth, in our solar system, in our galaxy, and beyond. As part of this quest, the Agency is committed to assuring the safety of space explorers: the humans who will touch down on the surface of the Moon, Mars, and other planets.

To shape an agenda to meet its goals, the Agency is considering several major new initiatives, outlined by the National Commission on Space and most recently by a NASA task force chaired by astronaut Sally K. Ride (1, 2):

- Mission to Planet Earth: an enterprise that would use space technology to study Earth systems on a global scale

- Exploration of the Solar System: a mission that would investigate a Main Belt asteroid and a comet, explore Saturn and its largest moon, Titan, and culminate in robotic surveys of Mars

- Outpost on the Moon: an effort that would draw upon the accomplishments of the Apollo Program and continuing research to establish a permanent human colony on the Moon

- Humans to Mars: a program that would employ the information collected by robotic missions to land Americans on Mars early in the next century and to establish an outpost on the planet within the following decade.

The most current statements about the Nation's future in space include the National Space Policy, issued by President Ronald Reagan in February 1988. In this document, the President reaffirmed the importance of missions, such as the permanently manned Space Station, that would maintain the Nation's preeminence in space research and prepare a basis for the expansion of human activity into the solar system.

The Space Station, scheduled to begin operations in the mid-1990's, and all the future initiatives under consideration by NASA have important life sciences components. A number of Agency programs work to resolve the life sciences issues central to these missions. NASA's Life Sciences Division is responsible for many of these programs, which take their lead from the larger organization and conduct research into issues related to the safety of human space flight and into the origin, evolution, and distribution of life in the universe. The research programs are grouped into two basic areas: Space Medicine and Biology, and Biological Systems. Flight Programs, a third area in the Division, is responsible for facilitating the conduct of Life Sciences research in space.

Space Medicine and Biology

The overall tasks of Space Medicine and Biology are to assure the health and safety of U.S. space crews and to understand the biological effects of space flight on organisms. Four factors that may significantly limit mission duration are physiological deconditioning, the biological effects of ionizing radiation, potential psychological difficulties on the part of the space crew, and environmental requirements, including the need of life support on missions of extended duration. Each factor has many unknowns, and flight personnel may exhibit marked variability in their susceptibilities to serious or limiting damage.

Physiological deconditioning during space flight is a significant concern because its effects are not well understood. Soviet cosmonaut Yuri Romanenko returned to Earth on December 29, 1987, after a record 326 days in space. His return, covered in the world press, suggested that humans can exist for considerable periods in space and successfully readapt to conditions on Earth. However, the experience of one man, particularly a veteran cosmonaut, cannot alone provide a sufficient basis for broad and highly optimistic conclusions. Moreover, hard data are not generally available concerning Romanenko's physiological and psychological condition during his mission and following his return to Earth.

Space Medicine and Biology is responsible in part for ensuring that American crews can maintain acceptable levels of physiological conditioning throughout extended space missions, ranging from the 1 to 3 years that could be required for a round trip to Mars. Before the U.S. embarks on such missions, research needs to be conducted to determine the limits of human endurance in space, to find the physiological point of no return for space crews, to assess, prior to launch, the susceptibility of individuals to the various stresses, and to develop effective countermeasures to the effects.

Experiments conducted on the ground and in space have identified a few of the physiological effects of microgravity on humans, which relate to the redistribution and reduction in blood volume, muscle atrophy and calcium loss, and disturbances (including space motion sickness) caused by confusing sensory signals. This research has also tested countermeasures that will be useful on missions lasting many months. Information has not been collected, however, on the effects of longer exposures to microgravity. One of the greatest challenges for life sciences research is to develop countermeasures to muscle loss and bone demineralization, which will probably be continuous during prolonged exposures and may not be completely reversible upon the astronaut's return to Earth. Such research must be conducted in space.

Ionizing radiation poses some significant questions for extended human missions in space. Considerable information is available about radiation beyond the protective influence of the Earth's magnetic field but little concerning the biological effects of HZE (high atomic number, Z, and high energy, E) particles or the shielding needed to protect crews on interplanetary missions from galactic cosmic radiation and solar particle events.

The interpersonal interactions among the space crew and between the space and ground crews will also be central to the success of extended missions. It will be particularly challenging to develop means to enhance productive behavior and avoid damaging conflicts since the complexities are great for a crew in a confined vehicle on an extended mission. The field of space psychology is still in early development. Information is not available on the levels of morale and performance possible for crews in space for lengthy periods of time or on the specific measures and systems needed to maintain these levels.

In its effort to understand the biological effects of space flight on organisms, Space Medicine and Biology concentrates on a fact simple in statement but complex in meaning: Gravity shapes life on Earth. We know that gravity has a role in determining the structure and function of terrestrial animals and plants. We do not know, however, the scope of this phenomenon, or the mechanisms by which gravity influences organism structure and function, or the mechanisms by which organisms can adapt to changes in gravitational fields. Indeed, we do not know if gravity is necessary for the maintenance of life. Space research with centrifuges — using short-radius capabilities for plants and animals and large-radii facilities for human studies — will make it possible to isolate the effects of gravity from other physiological changes. The data generated will help determine if life as we know it is possible in microgravity or at variable levels of artificial gravity. Extended human space flight into the inner solar system cannot be undertaken without this information.

Space Medicine and Biology covers not only the topics noted above, but also the requirements for a health maintenance facility, medical emergency measures, and a crew emergency return vehicle. The multiple challenges are detailed in the first six topical discussions given in section 3: "Biomedical Research," "Radiation," "Crew Factors," "Systems Engineering," "Operational Medicine," and "Gravitational Biology."

Biological Systems Research

This area incorporates programs in Controlled Ecological Life Support Systems (CELSS), Biospherics Research, and Exobiology. The central questions in these fields, a selection of which is given below, relate to the fundamental nature and limitations of life in the universe and require access to space for their resolution.

- Controlled Ecological Life Support Systems

  — What are the effects of weightlessness and space radiation on plants?

  — Can closed ecological systems be engineered to produce adequate food and to recycle wastes for extended space travel and settlement on other planets?

- Biospherics Research

  — What maintains the stability of the Earth's global ecosystem?

  — How are human activities disturbing that stability, and what can be done to preserve the health of our planet?

— What regional and global observations are required to assess the present condition and to predict future states of the world's ecosystems?

- Exobiology

    — What factors are required for the generation and evolution of life, and are these factors unique to Earth?

    — Did life ever evolve on Mars and, if so, what happened to this life and what are the implications for life on Earth?

    — How do astrophysical processes — such as solar activity and comet or asteroid impacts — influence the distribution of life on evolving, habitable planets in the cosmos?

The goal of CELSS research is a system that regenerates food, air, and water for crews on long-duration space flights. Several technologies may be considered for recycling air and water in a closed system, but biological processes must synthesize the complex materials needed to sustain human life. The CELSS Program has begun testing "Breadboard," a pilot-scale biomass production chamber designed to help develop a bioregenerative life support system. The research associated with this effort and related programs extends from investigations of plant photosynthetic processes to the physics and chemistry of supercritical wet oxidation of wastes. The Breadboard Project will provide essential information on the stability and reliability of bioregenerative life support systems. It remains an open question whether such systems will flourish in space.

Activities of the Biospherics Research Program are central to the Mission to Planet Earth. They proceed from the recognition that biological processes have shaped the chemical history of this planet. Interactions between the atmosphere and biosphere vary over time, and the record of these changes preserved in sedimentary rocks has been studied intensively to gain an understanding of the origin and evolution of life on Earth. Recently, the pace of change has accelerated. Human activities, including fossil fuel combustion, land use changes, and applications of novel industrial chemicals, have increased the concentrations of greenhouse gases and other atmospheric constituents markedly. A workable, descriptive theory of the biosphere is needed to understand the causes and consequences of these alterations. Environmental and biological data must be collected on an unprecedented worldwide scale to provide the basis for developing such a theory. Space capabilities are essential to this effort because of the global view they afford and their increasing ability to sense surface and atmospheric conditions remotely.

The questions posed by exobiologists are scientific iterations of queries long pondered by humankind, such as: Are we alone in the universe? How and where did life begin? Robotic probes followed by human missions to Mars, which is near the limits of our flight technology, will provide unique opportunities to obtain some answers.

Current knowledge suggests that water, a prime requisite for life, once coursed across the surface of Mars and that the early environment on the planet was similar to Earth's when life arose. Because the record of environmental conditions on Mars during its first billion years is potentially far better preserved than that of early Earth, samples from the planet collected by automated reconnaissance could fill the gap in Earth's geological record.

The *Viking* spacecraft, which landed on Mars in 1976 and transmitted data from the planet to Earth until 1982, showed no evidence of life or organic matter at two landing sites. This information, however, is not necessarily representative of the planet as a whole, and it does not address the possible existence of fossil organisms. Some indication of the former presence of life may be obtained by machines. Robotic surface reconnaissance could survey terrain where water may have existed in the past. In the process, it could probably identify strata of limestone or other minerals and organic compounds that are associated with biological activity on Earth and, in addition, possibly provide an early indication that Mars once harbored life.

Any valid indication of life on Mars would be a major scientific discovery. It would confirm the perception of many exobiologists that life is a nearly inevitable consequence of chemical evolution on any planet where environmental conditions are favorable, and it would have large implications for future research.

There is a fundamental human urge to know who we are, how we came to exist, what our place is in the universe, whether we can live elsewhere in the solar system, if we are alone. The scientific inquiry conducted by NASA Life Sciences programs into these questions is considered further in the following parts of section 3: "Controlled Ecological Life Support Systems," "Biospherics Research," and "Exobiology."

Flight Programs

The responsibilities of Flight Programs are to develop the equipment, facilities, expertise, and flight opportunities needed to assure successful conduct of life sciences investigations in space, to transfer knowledge gained from space flight to the larger research community, and to develop new technologies and equipment for future research conducted on the ground and in space. Its greatest current task is to see that a sufficient number and variety of flight opportunities are made available for life sciences investigations.

During the first half of the 1980's, Flight Programs concentrated on life sciences research for the Space Shuttle. An extensive inventory of laboratory equipment was developed, including controlled habitats for plants and animals and medical laboratory facilities for the study of humans in space. This equipment offered the flexibility necessary for various classes of experiments, such as small, self-contained studies, research using minilabs, and investigations requiring dedicated Spacelab missions.

The *Challenger* accident interrupted plans for experimentation using the Shuttle. This suspension was a serious blow to life sciences researchers, many of whom have waited 10 years or more to fly their experiments. Competition among these scientists and others will be intense for research space when Shuttle operations resume on a regular basis.

The achievement of life sciences' goals requires experimentation in space: on all manned missions, on Earth observing satellites, on orbiting observatories, on solar system explorations, on other planets. From past and continuing efforts, we know that the scientific rewards will be substantial, both for basic research and for future NASA initiatives. We realize, too, that flight opportunities in the foreseeable future will continue to be costly and limited in number; demand will exceed supply. The Life Sciences Division will have to work in this environment; the various programs and offices will need to achieve maximum scientific returns from available opportunities, including agreements with domestic and international organizations. Division research will continue to require flexibility and, most certainly, collaboration.

The issues outlined above are examined further in the remaining topical discussions presented in section 3: "Flight Programs," which focuses on requirements for flight opportunities; "Program Administration," which explores the management structure of NASA's Life Sciences Division and related programs; and "Applications," which examines the transfer of NASA's technological innovations to the private sector.

Future Course

The U.S. space program has reached an important threshold. In the past, NASA had to concentrate its funds on engineering and technical issues to make missions feasible at the most basic levels. NASA is now considering increasingly complex missions, looking both to intensified satellite observation of the Earth and extended human exploration of the inner solar system. Life Sciences programs must be positioned to help the Agency prepare for and conduct its future missions.

The research goals and emphases of the Life Sciences programs are truly diverse. This diversity springs from a number of sources. The Agency's needs are diverse, requiring research in basic science and in human health and safety. Modern science is generally progressing on a path that promotes interdisciplinary research. More specifically, space flight has required interdisciplinary science from its beginnings. As indicated by fundamental ecology, diversity can lead to synergisms and creative new possibilities.

The reports of several task forces comprised of prominent scientists and engineers have recently emphasized the importance of life sciences to the Nation's future in space (1, 2, 3). Their conclusions provide support to the findings and recommendations of the Life Sciences Strategic Planning Study Committee. The message from these various studies is clear and the opportunity at hand. **To conduct any of its current initiatives, to reassert its leadership in space research and exploration, NASA needs to assure that the life sciences are a critical part of the Nation's space program.**

Section 2 of this report presents the major findings and recommendations of the Life Sciences Strategic Planning Study Committee. These findings and recommendations were derived from summaries of topical studies conducted by the Committee, given in section 3.

Reference List

1. National Commission on Space. Thomas O. Paine, Task Group Chairperson. May 1986. *Pioneering the Space Frontier.* New York: Bantam Books.

2. National Aeronautics and Space Administration. Office of Exploration. Sally K. Ride, Task Group Chairperson. August 1987. *Leadership and America's Future in Space.* Washington, DC: NASA.

3. National Academy of Sciences. Committee on Space Biology and Medicine. Jay M. Goldberg, Committee Chairperson. 1987. *A Strategy for Space Biology and Medical Science for the 1980s and 1990s.* Washington, DC: National Academy Press.

2. Findings and Recommendations

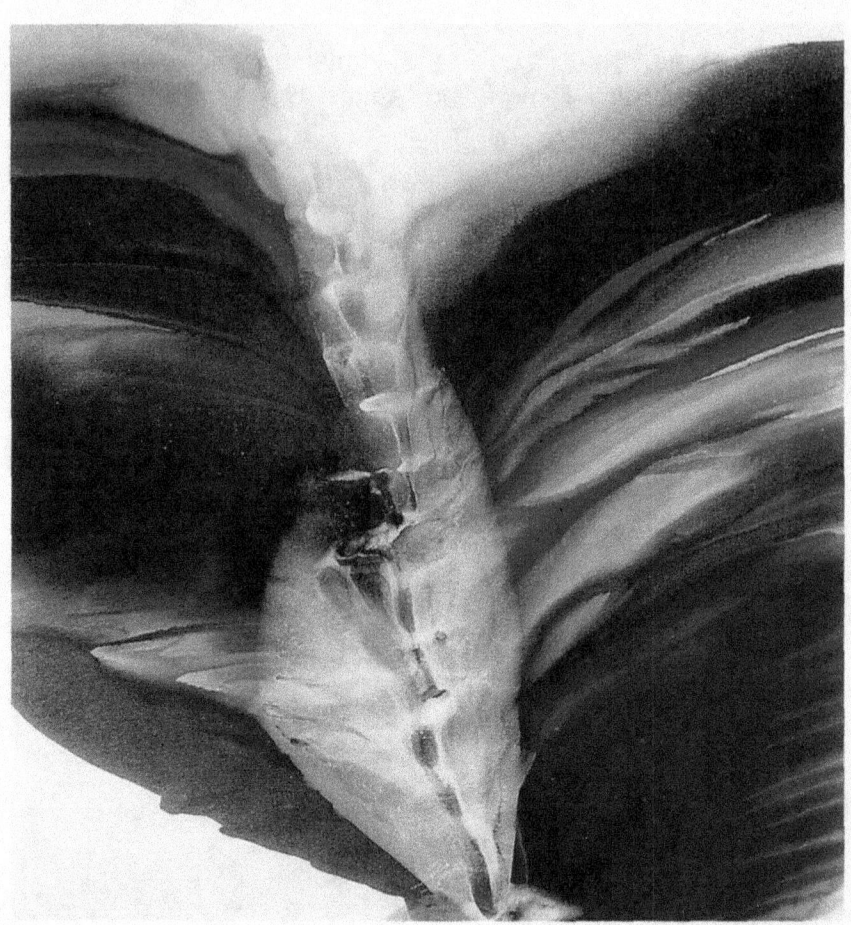

Phenomena Saturn Burns, 1974
© Paul Jenkins

2. Findings and Recommendations

This section is the central part of the Life Sciences Strategic Planning Study Committee (LSSPSC) report, for it highlights the Committee's overarching recommendations, strategic milestones for achieving those recommendations and findings and recommendations related to particular subject areas. The material emerged from the LSSPSC Study Group reports given in section 3, which present corresponding and more detailed findings and recommendations. It is organized in the categories itemized below, incorporating the subjects explored by the Study Groups:

- **Human Space Flight** focuses on the physiological and psychological challenges to humans in space and on the research and facilities necessary to overcome factors that may limit the success of manned missions, especially of extended duration.
- **Gravitational Biology** is concerned with the influence of gravity on the structure, development, and function of plants and animals.
- **Planetary Biosciences Research** concentrates on scientific issues pertinent to the origin, evolution, and distribution of life in the universe and the relationship of a planet's biota to its biosphere.
- **Flight Programs** emphasizes the need for flight opportunities for life sciences research, including dedication of Space Station laboratories for clinical and basic biological research.
- **Program Administration** itemizes administrative and organizational issues important to strengthening the work of NASA in the life sciences.

Overarching Recommendations and Strategies

In developing their summary papers, the LSSPSC Study Groups came to a number of parallel conclusions about life sciences at NASA and devised several similar recommendations. The LSSPSC determined that these recommendations were basic to the success of the Life Sciences program and, by extension, to the achievement of NASA's overall goals and long-range strategies, particularly as they affect human exploration of the solar system. The Committee presents these recommendations in the box on the next page.

The LSSPSC devised strategic milestones for fulfilling the requirements that are part of the overarching recommendations. These milestones, itemized according to 3-, 5-, and 15-year periods, emphasize the need to initiate work immediately, in the 1989 fiscal year.

Overarching Recommendations

To resolve life sciences issues critical to the success of the civilian space program, NASA should:

- Maintain and expand the Nation's life sciences research facilities located at the Agency's field centers, universities, and industrial centers by:
  - Establishing a mechanism for attracting promising young scientists to work on NASA projects and developing additional training programs at major universities and appropriate NASA installations
  - Establishing a program of NASA-supported professorships in space life sciences at selected universities
  - Encouraging industries to develop capabilities in space life sciences through technology research and development.

- Assure timely and sustained access to space flight, thereby facilitating the conduct of critical life sciences experiments. This should be accomplished through:
  - Accumulating state-of-the-art instrumentation
  - Flying an augmented series of Spacelab missions
  - Using a series of autonomous bioplatforms to study radiation and variable-gravity effects
  - Dedicating suitable facilities on the Phase 1 Space Station complex for life sciences research
  - Conducting a major augmentation of life sciences capabilities during the early Post-Phase 1 period.

- Synergize the presently independent research activities of national and international organizations through the development of cooperative programs in the life sciences at NASA and university laboratories.

- Complete and consolidate the unique national data base consisting of basic life sciences information and the results of biomedical studies of astronauts conducted on a longitudinal basis. This data base should be expanded to incorporate information obtained by other spacefaring nations and be available to all participating partners.

Strategic Milestones for 1989-1991

Life sciences research requires replication to verify experimental results, a process that involves considerable time in planning and conducting the investigations, as well as in developing advanced technology. Working from this understanding, the Committee recommends that NASA should do the following in the next 3 years:

- **Strengthen the planning process of the Life Sciences Division by assuring its timely integration into the Agency's overall strategic planning process.**
- **Augment life sciences research programs to establish the base of scientific knowledge required by planners and engineers to conduct missions relevant to Agency goals.**
- **Provide adequate funding to develop new state-of-the-art flight hardware for upcoming manned and unmanned life sciences missions in space. Such an investment will have a significant impact on the field of biomedicine not unlike the impact of the Apollo Program on medicine and space science.**
- **Initiate advanced technology development in the areas of minimally invasive biomedical instrumentation, biological remote sensing, exobiological flight instrumentation, and microwave signal processing.**
- **Increase the frequency of life sciences data acquisition on the Space Shuttle and international missions.**
- **Conduct a study to determine the requirements for extravehicular activity (EVA) for the next 20 years, to delineate innovative options, and to identify needed technologies.**

Strategic Milestones for 1989-1994

The next milestones are gauged for completion of life sciences preparations for the Space Station, scheduled to begin operations in the mid-1990's, and to implement a project requiring immediate action. For 1989-1994, the LSSPSC urges the Agency to:

- **Operate reusable biosatellites to obtain environmental, radiation, and artificial variable-gravity data on plants and animals.**
- **Achieve ground-based validation of major physiological and psychobiological countermeasures for long-duration missions.**
- **Conduct ground-based research on bioregenerative life support systems to achieve 90-percent closure.**
- **Initiate the Microwave Observing Project of the Search for Extraterrestrial Intelligence (SETI) Program.**

Strategic Milestones for 1989-2004

The 15-year plan looks to missions beyond the Space Station and asks the Agency to:

- **Establish a combined national and international life sciences research facility on the Space Station. This facility must support basic research on plants,** animals, and humans necessary to develop an understanding of the fundamental biological processes affected by gravitational forces.

- Develop an advanced biomedical research facility in space to investigate and verify technologies and medical support necessary to enable the planning and implementation of human exploration of the solar system.

- Develop and test in space a fully operational bioregenerative life support system(s) for future use in solar system exploration.

- Conduct cooperative missions with other national and international organizations to study the behavior of the biosphere and the origin, evolution, and distribution of life on Earth and in space.

The following subdivisions of section 2 present more detailed findings and recommendations relevant to particular subject areas.

Human Space Flight

Most of the major initiatives being considered by NASA involve human space missions of increasing duration — from 180-day rotations on the Space Station, to several months at a possible lunar colony, to 1 to 3 years on a round trip to Mars. For such missions to be possible, NASA's manned space flight program must undergo a decisive transition by the end of this century, surmounting significant problems in biomedicine, technology, and flight operations. The findings itemized below identify the primary challenges involved with human space missions of extended duration. The recommendations indicate the types of ground-based and space research that must be undertaken to resolve the outstanding issues.

Findings

- Four challenges potentially limit the duration of human space flight:
  - Physiological deconditioning
  - The biological effects of exposure to ionizing radiation
  - Potential psychological difficulties on the part of the space crew
  - Environmental requirements, including the need of life support on long space journeys.

- Ground-based experiments can provide significant data in the four areas. This research must, however, be validated and advanced by experimentation in space.

- Resolution of the concerns in each of the four key areas will require extensive research.
  - Zero gravity cannot be reproduced in ground-based research. Nevertheless, studies with human and animal models on the ground can provide insights into many of the physical effects of weightlessness, such as bone and muscle loss, cardiovascular deconditioning, and changes in fluid balances. Exceptions to this approach include neurovestibular effects and the loss of red blood cells, which require space research. In addition, the human research needed to vali-

date countermeasures against the deconditioning effects of weightlessness can only be done in space.

— Research on the effects of radiation on humans in space must proceed along two fronts. Characterization of the radiation fields, such as solar particle radiation and galactic cosmic radiation, is essential for predicting the specific risks and results of irradiation. To a large extent, this work must be conducted in space and may be done safely using experimental plants and animals. The space radiation environment is unique, and the spectrum of biological effects is not yet fully understood. Some of the questions may be studied on the ground using recently developed accelerators, but space-based experiments remain essential.

— For the rest of this century, ground-based research will be a practical mode for controlled experiments on group behavior and for developing methods to enhance crew performance on extended space missions. The effort will be particularly challenging, since group psychology pertinent to space flight is still in an early stage of development.

— Efficient Space Station operations and long-term human space flight will require substantial developments in life support and EVA technology and in the design of environments and systems to support crew health, safety, and performance. At present, however, efforts in these areas are fragmented among several program offices.

- A variable-gravity facility is a necessary tool for research conducted on the Space Station.

— It would provide a control for experimentation. Such a facility would, for example, supply the flexibility necessary for studies in space of the physiological effects produced by exposure to weightlessness and varying levels of artificial gravity.

— It would also allow testing of artificial gravity as a possible countermeasure to the physical deconditioning caused by extended exposure to weightlessness. Studies could be conducted with laboratory animals to generate data for potential human application.

- Space experiments to evaluate stay times for the Space Station crew and prepare for long-term human space missions will require Space Station laboratories for clinical and basic biological research as soon as manned operations begin.

- Provision of medical care for the crew figures prominently into plans for human space flight. Topics of consideration include the types of medical training required for crew members, the data needed for a medical information system, and the capabilities desirable in a Health Maintenance Facility (HMF) and a Crew Emergency Return Vehicle (CERV).

- Systems for storage, retrieval, and analysis of NASA's mission-derived data have been used as tools in the physical sciences for several years. Notable among these are the Climate Data System and the Pilot Land Data System developed at NASA's Goddard Space Flight Center. Life sciences programs have the beginnings of a data system in the astronaut medical information data base. At present, however, no standardized and formal system exists for archiving and analyzing the information derived from NASA's life sciences missions. Without such a system, and without even a directory of available data, valuable information about these missions may be overlooked or lost.

- Many scientific agencies in the United States and abroad have research interests parallel to NASA's in physical conditioning, radiation tolerance, interactions among crew members, and life support requirements. NASA and the other agencies could benefit by enhancing cooperative research, beginning with ground-based models and continuing with experimentation in space.

Recommendations

- NASA should immediately expand its program of ground-based research to resolve the outstanding questions about physiological deconditioning, radiation exposure, potential psychological difficulties, and life support requirements that may limit stay times for personnel on the Space Station and more extended missions.

  — Research should focus in part on the type of Space Station program needed to validate the models used and test the countermeasures developed in the ground-based program.

  — Comparability should be achieved between ground-based and space-based data by maintaining an atmospheric composition equivalent to that on Earth inside the pressurized module of the Space Station.

- NASA should plan an orderly, phased introduction of advanced life support and EVA technology into future manned space systems.

  — As part of this effort, the Controlled Ecological Life Support Systems (CELSS) Program should conduct experiments to determine whether reduced gravity levels are sources of stress that make plants less productive.

  — These experiments should be conducted in cooperation with the Gravitational Biology Program.

- NASA should design and build a suite of variable-gravity facilities for life sciences research.

  — They should be of sufficient size to accommodate various plant and animal specimens for basic research in gravitational biology and to test centrifugal fields as a countermeasure to microgravity in laboratory animals.

  — In addition, these facilities should evolve to include a human-rated system.

- In allocating payload and support resources for the Space Station, NASA should give first priority to life sciences research that will make human missions of extended duration possible. Laboratories for clinical and basic biological research should be available as soon as manned operations begin.

- NASA should take the following steps to ensure crew health and safety on the Space Station and missions of longer duration:

- Include a physician among the crew and train all other crew members to respond to medical emergencies. The physician should meet requirements established by the Operational Medicine Program and be trained as well to contribute to multidisciplinary mission objectives.

- Pursue longitudinal studies to collect biomedical data on all astronauts and an age-matched control population. This information, along with pre-, post-, and inflight data, should be integrated into a medical information system.

- Develop a Crew Emergency Return Vehicle to allow transport of crew members to Earth in case of space system and/or medical contingencies, as well as possible disruption in services provided by the Space Transportation System.

- Develop the capabilities of the Health Maintenance Facility with the ultimate goal of achieving autonomy for a Mars mission.

- Give priority to testing medical technologies necessary for the success of long-duration missions.

- Continue to recognize the Medicine Policy Board as the Agency's highest authority on issues of crew health and safety.

- NASA's Life Sciences Division should expand the existing data base of astronaut medical information to include a data base for all life sciences missions in space. The data base should take two forms:

  - First, an index data base should be created to catalog all relevant life sciences data sets. The index should provide browse facilities and summary information to help NASA investigators find archives of life sciences data sets germane to their areas of interest.

  - A second data system should be created that provides a formal and standard archive for all past, current, and future life sciences mission information (both from U.S. and international flights) and that allows for data retrieval and analysis.

- In conducting ground-based and space research in the life sciences, NASA should identify other scientific agencies in the U.S. and abroad that have parallel interests and should work actively to secure their collaboration in joint projects.

Gravitational Biology

Gravitational Biology focuses on the role of gravity in the reproductive, developmental, and metabolic activities of all forms of life. It is one of the few NASA programs that has both an intrinsic need for microgravity and the potential to make important contributions to basic science, as well as to operational research.

Findings

- Access to microgravity in space is crucial to developing an understanding of the role(s) of gravity in biologic processes.

- Extended periods of microgravity cannot be simulated on Earth; space flight provides the unique opportunity for investigating the effect(s) of microgravity on biologic systems.

- To date, opportunities provided by NASA for inflight life sciences research have been wholly insufficient.

• The success of long-term human activities in space will depend on a fundamental understanding of the effect(s) of gravity and especially microgravity on the metabolism, developmental biology, and life cycles of plants and animals. To develop this understanding, the Gravitational Biology Program will need to conduct long-term experiments in space and on the ground involving a large number and variety of research specimens.

• Conclusions from space-based research are not valid unless verified by adequate control.

- A variable-gravity centrifuge facility can provide the experimental control needed to isolate the effects of microgravity from all others encountered in space flight, such as solar and cosmic radiation, launch and reentry forces.

- In addition, variable-gravity centrifuge facilities will be needed to help test countermeasures to space flight deconditioning, to understand readaptations to 1 gravity, and to investigate a host of other phenomena of interest to clinical and biological investigators.

• Gravitational Biology can make crucial contributions both to basic and operational research programs.

- Research in Gravitational Biology is markedly multidisciplinary and is intimately interrelated with efforts in other areas of space life sciences research — most notably, the CELSS and Biomedical Research Programs.

- This synergism, critical to successful accomplishment of NASA's overriding goal of allowing humans to maintain a permanent presence in space, can continue and expand only if NASA provides the required facilities, funds, and personnel.

• NASA does not adequately support academic programs in Gravitational Biology. Because of limited support and few flight opportunities, research often requires more than a decade for completion.

- This situation deters students from posing questions that necessitate inflight experiments and effectively discourages them from pursuing studies in this area of science.

- The policies that have created the current situation will have to be changed if the United States is to regain leadership in the long-term human exploration of space.

Recommendations

• **NASA should increase the number and duration of life sciences experiments flown in space. These experiments should be conducted on a regular and fre-**

quent basis, with followup experiments flown in a timely fashion.

- Adequate inflight research capabilities must be provided, including variable-gravity facilities, on-orbit analytical equipment, and plant and animal vivaria capable of supporting successive generations subjected to varying, controlled gravity levels.

- Gravitational Biology research should be coordinated with that conducted by interrelated science programs, such as CELSS and Space Biomedicine. Resources, data, and personnel should be managed to allow a free flow of information among the various research projects and to enhance their relevance to the Nation's space program.

- NASA should operate its intramural and extramural research programs in a manner that attracts and supports excellent new researchers, especially young scientists, into the relatively new field of Gravitational Biology, as well as into other areas of space life sciences.

Planetary Biosciences Research

This section presents findings and recommendations pertinent to the Biospherics Research and Exobiology Programs, both of which currently depend on other Office of Space Science and Applications (OSSA) divisions for opportunities to conduct research in space. In addition, both have broad scientific charters, focusing on biological processes that operate from local to planetary levels. The findings and recommendations given below address organizational matters first, followed by scientific issues.

Findings

- The Biospherics Research and Exobiology Programs are developing plans for cooperative research with other OSSA programs having similar interests. Joint programs are being formalized as follows:

  — Between the Biospherics Research Program and the Terrestrial Ecosystems Program in the Earth Science and Applications Division

  — Between the Exobiology Program and the Planetary Exploration Program in the Solar System Exploration Division.

- Biospherics Research and Exobiology depend for space flight opportunities on missions sponsored by other OSSA divisions.

  — Data for the Biospherics Research Program will come from the types of Earth orbital missions to be included in the Earth Science and Applications Division's initiative for the Earth Observing System (EOS) (see Findings for Biospherics Research, section 3).

  — Planetary data for the Exobiology Program will come initially from automated missions sponsored by the Solar System Exploration Division, although manned Mars missions will be a major data source later.

- Limited resources in the Biospherics Research and Exobiology Programs have generally constrained the development of advanced sensing techniques and new methods for integrating sensor data. Such technology development is essential in enabling these programs to participate fully in research opportunities provided by NASA missions in air and space.

- The Life Sciences Division has the ability to undertake a comprehensive search for extraterrestrial intelligence, an effort that will have strong public appeal, allow for broad international cooperation, and make unique scientific contributions to radioastronomy.

- The Life Sciences Division has a particular interest in Mars, which occupies a unique position among the planets of the solar system.
  - Recent scientific evidence indicates that sometime during the first billion years of its history Mars was remarkably similar to early Earth with respect to the presence of liquid water, a volatiles-rich atmosphere, and a warm climate.
  - Life emerged on Earth during this same period and, according to current theory, may also have developed on Mars.
  - At present, the technology for robotic round trip and sample return has not been developed.

- Recent studies by committees of the NASA Advisory Council and the National Academy of Sciences (NAS) articulated an extensive list of research objectives and accompanying scientific strategies for the disciplines of biospherics and exobiology.
  - The LSSPSC concurs with the findings regarding global-scale Earth studies in *Earth System Science: A Closer View* (NASA, 1987) and *Global Change in the Geosphere, Biosphere* (NAS, 1986). We also agree with the recommendations for exobiology proposed by the NAS Committee on Planetary Biology and Chemical Evolution.
  - Present funding constrains the Biospherics Research and Exobiology Programs from exploiting the research opportunities delineated by the NASA and NAS committees.

Recommendations

- **NASA should make the science requirements of biospherics and exobiology integral to plans for its Mission to Planet Earth and Exploration of the Solar System initiatives.**

- **NASA divisions with similar interests in planetary biology — the Life Sciences, Solar System Exploration, Earth Science and Applications, and Astrophysics Divisions — should develop additional programs to promote maximum return from collaborative research.**

- **The Biospherics Research Program should participate in the development and implementation of the Earth Observing System.**

- **The Biospherics Research and Exobiology Programs should intensify development of the technology necessary to generate advanced systems for instrument** analyses, remote sensing, and data analysis. These systems will be essential in realizing the potential of scientific returns from missions involving Earth observation and exploration of Mars, the outer planets, Titan, comets, asteroids, and the Moon.

- NASA should initiate the Microwave Observing Project (Search for Extraterrestrial Intelligence) now, before radiofrequency interference makes it exceedingly difficult or impossible to conduct the research from Earth.

- NASA should pursue vigorous programs of ground-based research, remote observations, and instrument development for use on missions to assess evidence bearing on the possible origin of life on Mars and the nature of chemical evolution on other solar system bodies. Knowledge gained in this program will provide the scientific basis for future manned exploration of the planets.

- NASA should develop the technology of robotic round trip, sample selection and analysis, and sample return for exploration of the surface of Mars, asteroids, and comets. This effort should include precautions to avoid the spread of contamination within the solar system.

- NASA should significantly enhance the ground- and space-based research capabilities and infrastructure (funding, personnel, and facilities) for planetary biology in order to maintain the Agency's leadership role in planetary research, implement the science strategies recommended by the NASA and NAS advisory committees, and optimize the scientific return of future missions.

Flight Programs

The findings and recommendations given earlier in section 2, as well as those presented in the discussions of section 3, identify the need for a series of flight experiments designed specifically for life sciences research. The findings and recommendations listed below make a similar point. The purpose of this repetition is to emphasize the importance of increased flight opportunities for life sciences experimentation, which is the primary research requirement identified in this report.

Findings

- NASA has plans for advanced missions that will require long-duration space flight. Such missions include extended space travel in low Earth orbit on the Space Station and may ultimately involve missions requiring extended interplanetary travel, such as lunar colonies and voyages to Mars. Action is needed if NASA plans to validate extended stay times for Space Station crews and to preserve its options for piloted Mars missions.

  — The Agency's Life Sciences programs will play a central role in validating stay times and in certifying crews for its advanced missions.

  — Life sciences research will need to develop countermeasures to factors that may limit mission success, including physiological deconditioning, radiation hazards, and issues related to crew psychology and crew-machine interactions.

- Preflight delays and schedule instabilities for flights slated to carry life sciences experiments make it difficult for young scientists and graduate students to participate in life sciences flight experiments. The limitations on flight opportunities also pose difficulties for established investigators; many have waited 10 or more years to fly a single life sciences experiment.

- For life sciences experiments to provide a useful statistical base, they must be replicated under controlled conditions. This is true for flight experiments with humans, other animals, and plants, as well as for biospherics and exobiology experiments investigating biologic and biogenic phenomena.

Recommendations

- NASA should increase the flight opportunities for life sciences research associated with human space flight. Specifically, the Agency should:
  - Dedicate a greater number of regularly scheduled Shuttle middeck lockers and commercially developed flight facilities to life sciences experimentation.
  - Increase the flight rate (priority) of Spacelab and dedicate a larger percentage of Spacelab volume, time, and resources to life sciences issues.
  - Dedicate a clinical research center and a biological research center for life sciences experiments on the Phase 1 Space Station.
  - Deploy an unmanned spacecraft that is reusable and can support a variety of flight experiments, including those requiring a variable-gravity facility. The spacecraft should be designed for recovery and for rapid redeployment on an expendable launch vehicle.

- NASA should actively encourage students and non-NASA life scientists to participate in mission-related research but should be careful not to encourage unrealistic expectations of flight opportunities.
  - Announcements of Opportunity (AO's) should be targeted to a range of experimental opportunities available on the Space Shuttle middeck, Spacelab, free-fliers, Space Station, and on collaborative missions with other countries, such as the Soviet Union, the Federal Republic of Germany, and Japan.
  - AO's should be scheduled for release on a regular basis to give investigators the opportunity to plan their proposals and research programs.
  - Discipline Working Groups should be implemented to allow greater contact between investigators and the NASA programs where AO solicitations are initiated.

- A new generation of ground-based and flight-certified instrumentation should be developed to support the research objectives of the Life Sciences programs. This instrumentation should include the following:
  - Noninvasive monitoring techniques for biomedical applications

- Flight-certified, variable-gravity facilities on appropriate platforms to house plants and animals of various sizes and ultimately a human-rated, inflight, variable-gravity facility

- A capability for remote data collection, analysis, cataloging, and storage of biologic and exobiological data

- A capability for data-base management and data analyses of biomedical, biologic, and exobiological information.

Program Administration

The coordination of life sciences activities at NASA is a challenging task. The research is multidisciplinary in approach and involves many other organizations — both within and external to the Agency — that are pursuing similar interests. The findings and recommendations given below identify the administrative challenges, acknowledge recent progress, and specify resource requirements.

Findings

- During the course of this study, the life sciences have received increased attention within NASA.
  - Concern about the effects of long-duration space flight has given life sciences a higher priority in the Agency and has provided the program with an opportunity to articulate its own goals more clearly.
  - At the same time, however, senior managers have not always appreciated that life sciences concerns are unique in the study and maintenance of life in space and that this uniqueness creates special administrative challenges for the program.

- The Life Sciences Division does not have sufficient resources in funds, staff, and facilities to realize its own objectives or the objectives set for the program by senior managers.

- The dispersion of life sciences activities across a number of NASA program offices has made it difficult to conduct research in several important areas, particularly human factors and biospherics. While new coordination efforts are under way, the integration of life sciences efforts across the Agency remains problematic.

- NASA's Life Sciences Division supports diversified programs that could benefit from coordination between the Division and outside organizations. The Division has initiated formal cooperative agreements with the National Institutes of Health and other Federal agencies.

- The increasing importance of foreign space programs has opened up a broad field for potential cooperative projects. These arrangements require international negotiations that are lengthy and involve multiple U.S. agencies.

- The Life Sciences Division has not always been able to create stable relationships with outside scientific groups.

- Scientists outside the Agency provide a valuable resource to NASA, both as researchers and as advisors to Agency staff.
- Recent program development plans for a balance between external and intramural research, as well as the creation of a new advisory and planning structure, promise desirable change in this area.

• Information concerning life sciences activities is not disseminated as widely as possible and desirable. As a result, many university and industrial researchers find it difficult to secure data on past, current, and future life sciences projects.

Recommendations

• Senior NASA management should support the continuation of recent Division efforts to establish a strong program by:

— Strengthening the Division's role in Agency-wide planning

— Facilitating access to frequent and regular flight opportunities

— Acknowledging the differences between programs of the Life Sciences Division and other NASA program areas

— Indicating to the rest of the Agency that biomedical research relevant to the safe conduct of human space flight is essential to ongoing and future NASA initiatives.

• Senior personnel from the Life Sciences Division should participate in all top-level planning of Agency flight programs.

• NASA should substantially increase the resources for Life Sciences programs to assure implementation of the recommendations given in this report.

• NASA should increase its efforts to expand the numbers of scientists at the Centers and Headquarters and should institute new efforts to provide career development opportunities for existing staff.

• The Life Sciences Division should further its efforts to establish formalized agreements and working groups with other agencies and organizations.

• NASA should provide funds to expand and implement plans to establish Specialized Center of Research (SCOR) units within selected universities, an effort designed to develop young scientists in space life sciences.

• In addition, the Agency should consider the establishment of NASA-supported professorships in space life sciences at selected universities, so that by 1990 one or two internationally recognized bioscientists and clinical investigators can play a significant role in the biomedical research crucial to human space missions of extended duration.

• The Life Sciences Division should generate and maintain a data base through collaborative arrangements with NASA's Scientific and Technical Information Facility and the National Library of Medicine.

3. Life Sciences in the Space Program

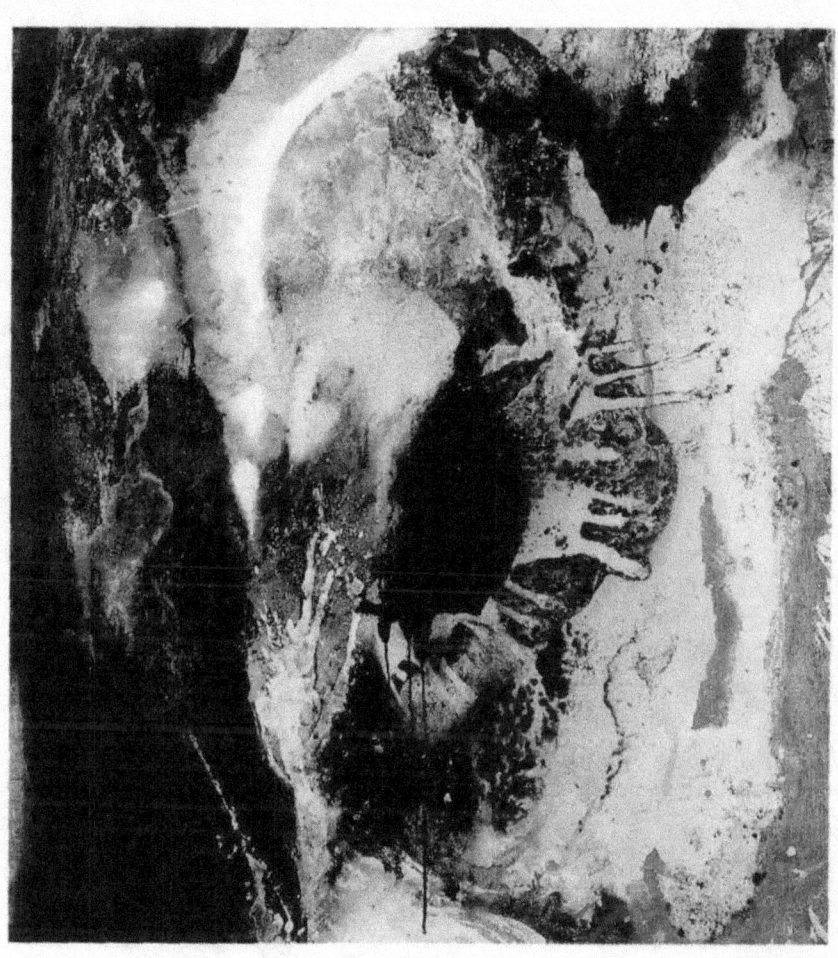

Shooting the Sun, 1956
© Paul Jenkins

3. Life Sciences in the Space Program

The Study Groups of the Life Sciences Strategic Planning Study Committee (LSSPSC) conducted indepth evaluations of NASA life sciences. They surveyed the scientific literature in their respective fields and interviewed researchers and administrators at various offices and divisions within NASA Headquarters and at the field centers, particularly at Ames Research Center (ARC) and Johnson Space Center (JSC). Some Committee members also visited the life sciences research facilities in the Soviet Union, Federal Republic of Germany, France, and England. In addition, the LSSPSC invited representatives from the European Space Agency (ESA) and the National Space Development Agency of Japan (NASDA) to participate in their initial meetings.

The Study Groups organized their material into summary papers, presented in this section as follows:

- Biomedical Research
- Radiation
- Crew Factors
- Systems Engineering
- Operational Medicine
- Gravitational Biology
- Controlled Ecological Life Support Systems (CELSS)
- Biospherics Research
- Exobiology
- Flight Programs
- Program Administration
- Applications.

The Study Group reports present overviews of the given topics. The heart of each document is the concluding list of findings and recommendations, which provided the substance for the LSSPSC's overall findings and recommendations.

Bernadine Healy, M.D.
Chairperson

William DeCampli, M.D., Ph.D.

Frederick C. Robbins, M.D.

Warren Lockette, M.D.
Staff Associate

Biomedical Research

The primary goal of NASA biomedical research is to ensure the safe transport of humans into space and back to Earth and the safe maintenance of humans living and working on a long-term basis in space. To achieve this goal, several NASA field centers are engaged in biomedical research. Intramural biomedical research is conducted primarily at three NASA laboratories located at Ames Research Center (ARC), Johnson Space Center (JSC), and Kennedy Space Center (KSC).

Issues and recommendations pertinent to the NASA Biomedical Research Program are summarized in this report. Information was collected from briefings given by investigators in the various disciplines at each center. Briefings were also provided by staff at NASA Headquarters, the U.S. Air Force (USAF) School of Aerospace Medicine and the Human Systems Division, the Naval Medical Research and Development Command, the Air Force Office of Scientific Research, and the National Institutes of Health, and by individual scientists funded through the extramural program in biomedical research at NASA. In addition, NASA publications were reviewed.

Scientific Issues

Numerous individual technical reports and a small number of reports in the peer-reviewed literature have reiterated the rationales for conducting biomedical research at NASA, provided a history of biomedical research at the Agency, and presented the findings of biomedical research sponsored by NASA (1,2,3,4,5,6). These and other references, including *A Strategy for Space Biology and Medical Science for the 1980s and 1990s* (7), describe the issues in biomedical research that must be resolved to ensure the safety of humans living in space. While the questions discussed below are important for short-term, Shuttle-type, and medium-duration (Space Station) missions, they become vitally significant for longer duration flights, such as an expedition to Mars or the development of a lunar base.

Cardiovascular Physiology

In microgravity, there is a loss of the gravity-induced vascular pooling of blood in the lower extremities that normally occurs in humans with upright posture. The volume of blood normally found in the lower extremities on Earth is centrally redistributed in the body when in microgravity. Volume and pressure receptors in the cardiovascular system sense the redistribution of blood volume and activate regulatory mechanisms that counteract this "central blood volume expansion" by reducing the total blood volume.

For example, the human body can decrease blood volume by increasing urine output, decreasing thirst, or changing the permeability of blood vessels to fluid. These changes may be mediated by humoral agents, such as atrial natriuretic factor, vasopressin, angiotensin, aldosterone, and catecholamines; and these physiologic changes may also be reflected in hematologic indices, such as the hematocrit (percent volume of red blood cells/total blood volume). Redistribution of blood flow, reduced blood volume, and reduced cardiac output on short-duration missions may contribute to the deconditioning observed in astronauts on their return to Earth. Upon reentry into full gravity, the vascular pooling of blood in the lower extremities results in a decrease in blood flow to the cerebral vasculature and may result in syncope (loss of consciousness) if appropriate countermeasures are not taken (4, 5, 6, 8, 9, 10, 11).

In addition, many astronauts experience space adaptation syndrome, characterized by nausea, emesis, nasal congestion, diaphoresis (sweating), and fatigue. Signs or symptoms of space motion sickness, which can be incapacitating, have been reported in a significant number of astronauts: 11 out of 33 Apollo astronauts, 5 out of 9 Skylab crew members, and 6 out of 12 early Shuttle crew members (9).

Astronaut Owen Garriott, at left, draws blood from fellow crew member Byron Lichtenberg during a biomedical research experiment aboard Spacelab 1.

Other cardiovascular concerns are related to space flight. Acute acceleration has been associated with cardiac dysrhythmias, particularly bradycardia. A broad range of arrhythmias, including both atrial and ventricular arrhythmias, has occurred among astronauts during space flight. It has been reported that a Soviet cosmonaut was recently rescued after 6 months in space because of cardiac dysrhythmias. Little information exists on the prevalence and pathophysiology of cardiac arrhythmias in space, a concern especially important for long-term missions.

Other aspects of cardiovascular deconditioning have been reported during space flight. These involve complex alterations in cardiovascular reflexes, endocrine status, and possibly cardiac function. However, reports of cardiac deconditioning are based on either a small number of echocardiographic data from Shuttle flights or responses to orthostatic stress immediately following space flight. Inflight echocardiographic data are consistent with changes in fluid compartmentalization, but it is unclear if myocardial performance (e.g., inotropism or chronotropism) is altered independent of changes in blood volumes.

To supplement the scant data gathered from space missions, the cardiovascular response to weightlessness has been studied by NASA using bed rest with head-down tilt to simulate the effects of weightlessness on hemodynamics. When an individual is placed in a head-down tilt for a number of days, it is hypothesized that the person will experience the expansion of central blood volumes in the thorax as the blood returned from the veins of the legs is increased. Bed-rest studies are well controlled, can be done on Earth, represent a continued line of investigations not dependent upon flight availability, and pose questions relevant to clinical concerns beyond the space program. Although this method may be effective in simulating exposure to microgravity, a significantly larger number of inflight studies must be done to compare the reliability of using bed-rest protocols to study the effects of prolonged exposure to microgravity (8,9). It cannot be overemphasized that bed rest (even with head-down tilt) at 1 gravity (g) on Earth does not mimic or reproduce the microgravity environment of space because gravitational forces are still at work. Despite the recumbent posture, gravitational forces still operate on the circulation, bones, and muscles. Such Earth-based experiments on humans are merely a small first step toward understanding microgravity deconditioning.

Several countermeasures for cardiovascular deconditioning have been investigated. These methods include the use of positive pressure suits, volume replacement with water and salt, pharmacologic agents to promote the retention of electrolytes and fluid, the use of applied lower body negative pressure, and exercise. However, these prophylactic measures have not been uniformly effective at reducing the incidence of near-syncope following Shuttle reentry. Accordingly, since cardiovascular deconditioning and space adaptation syndrome may decrease crew performance, and since the mechanisms causing these changes are poorly understood, further efforts are needed to define and prevent or treat these physiologic responses.

In summary, significant changes occur in the cardiovascular system in microgravity, and many questions remain unanswered. What is the role of the cardiovascular system in the etiology of deconditioning? Is space travel associated with an increase in morbidity from cardiovascular disease in flight crews? Does the degree of cardiovascular deconditioning from short-term space flights predict incapacitating cardiovascular deconditioning with longer flights? Do cardiovascular responses to microgravity detrimentally affect other organ systems? For example, do the hemodynamic and hormonal responses to microgravity result in alterations in vestibular function or cognitive function? Can improved countermeasures be developed for the problems that occur with space travel? What is the role of exercise in preventing the cardiovascular deconditioning associated with space travel?

A matter of terminology is significant. The word "adaptation" denotes a favorable modification of structure or function in response to environmental stress. The motion sickness of space adaptation syndrome often passes after the first day or two. This disappearance of motion sickness may be an adaptation to unusual sensory perceptions, much in the way a sailor adapts or gets sea legs after a few days at sea.

However, the word "deconditioning" signifies an unfavorable change. A well-trained athlete put to bed for a month undergoes deconditioning; a deliberate effort must be made to regain the effect of previous training. The cardiovascular deconditioning (and the changes in muscle and bone described below) adversely affects the astronaut upon return to Earth. We do not know if this deconditioning impairs the individual's ability to perform in space.

Neurophysiology and Behavioral Science

Among the sensory systems most likely instrumental in the pathogenesis of space adaptation syndrome, the vestibular system is the most probable candidate. The vestibular apparatus consists of semicircular canals in the inner ear that sense angular momentum and otoliths that sense rectilinear acceleration. The afferent and efferent neural pathways and neurotransmitters to and from the labyrinth to proprioceptive receptors, posterior columns, the cerebellum, and autonomic control centers in the medulla are not well understood. Currently, investigations of vestibular function in humans include measurement of Coriolis stress susceptibility (measuring an individual's susceptibility to motion sickness following movement out of the plane of rotation), and measurement of otolith function using acceleration sleds, swings, and parabolic flight. NASA currently flies a KC 135 aircraft in a parabolic profile to simulate microgravity; however, this does not provide a sustained environment of microgravity for more than 30 seconds, and the change in gravitational force is not uniform. Obviously, this technique has limited application, and short of actual flight time in space, a suitable mechanism does not exist to study vestibular function in microgravity.

Space motion sickness is one of the most serious concerns in short-duration space flight. The mechanism(s) is unknown by which labyrinth function in microgravity is altered. The sensory conflict theory, hypothesized to explain space motion sickness, postulates that "motion sickness occurs when patterns of sensory input to the brain from the vestibular system, other proprioceptors, and/or the visual system are markedly rearranged, at variance with each other, or differ substantially from expectations of stimulus relationships in a given environment" (9). In gravity, head movement is associated with changes sensed in both the otoliths and semicircular canals. In microgravity, there may be unexpected stimulation of only the semicircular canals. Whatever the cause, a large percentage of astronauts experience nausea, vomiting, and malaise. At present, the most effective pharmacologic treatment of space motion sickness is some combination of anticholinergic drugs (scopolamine) and amphetamine (11). Unfortunately, these drugs may be associated with a decrease in crew performance, and their effectiveness is unpredictable. Certainly, an understanding of the mechanism by which vestibular function changes in space will result in a more effective approach to the prevention and treatment of space motion sickness.

The scientific questions are clearly multidisciplinary. It is important to integrate the various research activities accordingly. For example, do the neurophysiologists investigating vestibular function consider that such changes may be brought about by the hemodynamic changes of microgravity being investigated by the cardiovascular physiologists?

This question and others have broad relevance outside the operational responsibilities of biomedical research at NASA. For example, does deterioration in vestibular function result from the hemodynamic changes associated with microgravity? Are these effects similar to those seen in Meniere's disease? The answers to these scientific inquiries and the solutions to these clinical problems may be found more expeditiously by close association between NASA and the biomedical research community external to NASA.

In short, the following questions in neurophysiology must be addressed: What are the mechanisms responsible for the changes in neural function that occur in microgravity? Do these changes in neural input contribute to the frequent reports of space motion sickness? Are microgravity-induced changes in neural function dependent upon the duration of microgravity, and what countermeasures will be successful to treat changes in sensory perception, postural control mechanisms, and neuroendocrine responses that occur in microgravity?

The effects of the isolation and microgravity incurred by long-duration space flight on interpersonal relationships, cognitive function, affect, and sexual function also need to be investigated. Previous studies (12) have been too brief to allow extrapolation for missions to Mars or the establishment of lunar bases.

Endocrine and Musculoskeletal Physiology

It has been reported that total body calcium losses average 0.3 percent per month during space flight, and it is believed that most of the calcium loss comes from weight-bearing bone (13). The loss of body stores of calcium may be due to decreased oral intake of calcium in space flight, decreased absorption of dietary calcium, increased calcium resorption from bone in microgravity, and increased urinary calcium loss. Serum calcium concentration is increased by parathyroid hormone, which also promotes urinary phosphate loss; ionized calcium concentration is decreased in response to calcitonin. Calcium levels are also affected by vitamin D and metabolites of vitamin D in the liver and kidneys. How the action of the hormones is affected by microgravity is not known.

The mechanism by which there is a net negative balance of calcium in microgravity is unknown. The low bone mass that results from increased calcium resorption of bone raises the theoretical concern of susceptibility to fracture. Also, hypercalciuria, or increased urinary excretion of calcium, may predispose individuals to nephrolithiasis (kidney stones), especially when phosphate excretion is increased. Although bone demineralization during weightlessness has not caused acute or chronic adverse health effects during or following space flight, the likelihood of such an eventual occurrence is not negligible when the heterogeneous renal transport kinetics that vary between individuals are considered. It is important to recall that much of the architecture of bone (examples are the cancellous bone struts in the femoral head and neck) results directly from gravitational stress acting on body weight, and that the reshaping of bone after fracture is closely related to the lines of weight-bearing force. The influence of gravity on both the macrostructure of the skeleton, and the microstructure of cortical and cancellous bone, is unquestioned.

Urinary excretion of calcium and phosphorus observed among the Apollo and Skylab crews paralleled the losses previously reported in healthy, immobilized bed-rest subjects on Earth (13). Urinary excretion of hydroxyproline and total and nonglycosylated hydroxylysine (indicators of bone matrix turnover) was also elevated in Skylab subjects. It is particularly troublesome that the continuous increase in calcium excretion during space flight showed no tendency to plateau. A conservative extrapolation of the amount of calcium lost during relatively short periods in space suggests that a 6-month mission would result in a loss of 2-3 percent of total body calcium (13). Measurements of bone density have corroborated the evidence of negative calcium balance from bone calcium loss, and there is added concern that this loss may not be recoverable after the flight and that this may result in less dense (i.e., weaker) bone in crew members subsequent to their missions.

Current evidence has not demonstrated an increase in morbidity or mortality from altered calcium metabolism after short space missions. However, the clinical effect of longer missions (greater than 90 days) on calcium metabolism and skeletal performance is unknown. Furthermore, although the obvious role of altered calcium metabolism on bone structure and function has taken priority in current studies, the data on hand suggest other concerns. For example, the negative calcium balance is associated with an increased loss in magnesium. Since hypomagnasemia is associated with altered coronary vascular reactivity and ventricular ectopy (14), it is quite possible that hypomagnasemia may increase the likelihood of cardiac dysrhythmias during acceleration or space flight.

In microgravity, the need to maintain skeletal muscle integrity is decreased since there is less need for active opposition to gravity to maintain posture or move limbs. Anthropometric measurements, stereometric analysis, and electromyographic data have demonstrated that with space flight, there is loss of muscle strength, a decrease in muscle mass, an increase in protein catabolism, and a persistently negative nitrogen balance (13).

Programs for prevention of muscle atrophy and skeletal demineralization are hampered by insufficient understanding of the metabolism of bone and muscle in space. Proposed countermeasures for muscle atrophy and skeletal demineralization include the use of exercise treadmills and cycle ergometers, electrical stimulation of muscle groups, and "Penguin suits," which oppose body movement and partially compensate for the lack of gravity on the antigravitational muscles. In addition to the loss of opposing force on antigravitational muscles, it remains to be determined what effect the hemodynamic changes (such as redistribution of blood flow) or endocrine changes (including calcium loss, negative nitrogen balance, or loss of potassium reservoirs) have in development of muscle atrophy associated with space flight. The impact of muscle loss on performance of astronauts in space remains unclear.

In summary, the questions that need to be addressed include the following: What is the mechanism of osteopenia (loss of bone tissue) that occurs upon exposure to microgravity? Is this osteopenia associated with an increase in crew morbidity, such as from nephrolithiasis (kidney stones)? What are the sequelae of osteopenia from short-duration flights? Will alterations in ionized calcium concentrations incapacitate crew members with cardiac dysrhythmias or pathologic fractures? What countermeasures may be developed to prevent osteopenia and the associated humoral changes? What countermeasures can be developed for the skeletal muscle atrophy associated with microgravity? What is the mechanism by which gravity or inertial forces retard skeletal muscle atrophy? Will osteoporosis of the bone of the ear affect auditory perceptions, much in the way Paget's disease may lead to hearing loss?

The effect of microgravity on reproductive function in space has been virtually ignored. Are there any effects of microgravity or cosmic radiation exposure during space flight on reproductive function or developmental biology in flight crew members?

As with cardiovascular research, the fields of neuroscience and endocrinology are likely to be advanced significantly as a result of space-based research. However, as with all scientific research, the appropriate controls must be done. Accordingly, the use of a space-based centrifuge must be considered.

The role of a human centrifuge in space, employed to abort the deconditioning of bone, muscle, and the cardiovascular system during space flight, has been an intriguing possibility for many years. Much more work needs to be done on centrifuge-simulated g forces and the "dose of centrifugation" required to abort deconditioning before a solid recommendation can be made regarding centrifuge therapy on prolonged space flights. Preliminary centrifuge studies of this type can be carried out in animals on either the Shuttle, or the Space Station, or both. A centrifuge large enough for humans, in space, could only be accommodated on a structure of approximately the size of the proposed Space Station module.

Hematology

A significant decrease in red cell mass has been reported in the Gemini, Apollo, Skylab, and *Soyuz* flight crews, and this decreased red cell mass cannot be explained by spacecraft hyperoxia. Increases in red cell destruction were not reported, and the downward trend of decreased erythropoietin levels studied in a few astronauts was not statistically significant (6). However, the expansion of plasma volume with saline and the use of antiemetic drugs confound the interpretation of the most recent results. No changes in immunoglobulin levels were reported in astronauts during Skylab; an impaired mitogenic response of lymphocytes to lectins has been reported (6).

These preliminary findings of anemia and decreased immune responsiveness following exposure to microgravity need verification. Pending the substantiation of these reports, the salient questions include: What is the etiology of anemia that has been associated with space flight? Are the preliminary findings of altered mitogenic responses of lymphocytes in microgravity clinically significant?

Logistic and Policy Issues

Having identified many of the major questions in biomedical research for a successful space program, we must determine the most effective methods of answering these questions. Many of the logistical and policy issues discussed below are not unique to biomedical research within the Life Sciences Division. These problem areas have been identified by the NASA Long Range Planning Committee for the Life Sciences, and they warrant special emphasis.

NASA Goals for Biomedical Research

Research priorities in the biomedical sciences cannot be based on the long-term goals of NASA, in large part because of the real or perceived lack of definition in these goals. Without a clearly defined national space goal, it is difficult to have an operational objective. A number of other reasons account for the uncertainty of research priorities at the Agency. Biomedical research at NASA is not a separate line item, and it is subject to variations in funding within the different offices. Funding of particular projects may also be subject to competition between the Centers and Headquarters.

Agency Problems in Attracting Quality Biomedical Researchers

The number of full-time employees in Life Sciences at NASA represents too small a percentage of the total number of Agency employees. Clearly, NASA has an insufficient number of top-level biomedical researchers. The reasons for this lack of manpower include the following: uncertainty exists concerning the importance of biomedical research at NASA; the time between award of grant and conduct of the experiment in space is too long, sometimes extending to 10 years and more; the paucity of data collected from space missions makes results difficult to interpret and publish in peer-reviewed literature; the thrust of NASA research seems operational in nature; opportunities are limited for interface with members of the scientific community external to NASA; the visibility of biomedical research at NASA is limited in the universities and industry because of a small extramural grant program; there is no effective way for senior NASA bioscientists to achieve the status of university-tenured faculty; it is difficult for individuals external to

NASA to become involved in biomedical research at the Agency; the Announcements of Opportunity and Requests for Proposals are not well placed; the peer-review process and external grant programs are not well understood by non-NASA investigators and the larger scientific community. For these and other reasons, NASA is not perceived as a place for young talent in the biomedical sciences to develop a research career. These factors also discourage senior scientists from seeking a place at the Agency. NASA needs to recruit and hire world-class scientists for its research programs.

Valuation of Biomedical Research at NASA and Other Organizations

Although biomedical research is not expected to be the prime mission of NASA, there seems to be only limited understanding of the essential role biomedical research will play in achieving a permanent human presence in space. The space flight missions of NASA and prolonged human space dwelling (0.5-3.0 years) cannot be achieved without a significant bioscience program in human and clinical research.

Biomedical research is not supported sufficiently by NASA. With a total budget in the Life Sciences Division never exceeding $70 million per year in the face of a $10 billion total NASA budget, it is difficult to believe that biomedicine and the other areas within the Division are a valued Agency component. But before one makes a plea for an increase in the budget, a focused and valued program must be endorsed and receive appropriate priority within NASA itself. It has also been suggested that given the previous funding levels, the expectations of biomedical research from NASA have been too great and the concerns for health safety advanced by biomedical researchers have been too cautious.

The need for a strong biomedical research program at NASA is also not clear to other agencies or organizations. NASA has done moderately well at advertising its technical accomplishments in engineering, but its accomplishments in the life sciences are not as well disseminated. A good mechanism does not exist for routinely determining the potential applications of this research at NASA to terrestrial-based problems or clinical medicine. Whereas most extramural researchers funded by the National Science Foundation are aware of NASA research, too few university and hospital-based biomedical researchers traditionally funded by the National Institutes of Health are familiar with NASA programs in biomedical research. It is possible that space-based research can advance terrestrial clinical science, but this likelihood is not well appreciated by life scientists unfamiliar with biomedical research at NASA.

Recommendations
Cardiovascular Physiology

- **Adequate numbers, verification, and control of experiments must be achieved if recommendations for countermeasures are to be made according to scientific merit.**

- Verification of the degree of cardiovascular deconditioning should be obtained while concomitant countermeasures are developed.
- Bed-rest studies and studies using lower body negative pressure should be continued, but they must be supplemental to inflight research.
- Instrumentation for onboard hemodynamic monitoring should be implemented according to a well-defined, long-term target.
- The role of exercise should be clearly defined in such areas as susceptibility to space deconditioning, prevention of cardiovascular deconditioning, and protection against cardiovascular dysfunction with prolonged space flight.
- Experimental studies should be conducted using humans and animal models.
- The use of a variable-gravity centrifuge in flight must be aggressively studied.
- Collaborative efforts should be encouraged between U.S. and Soviet scientists, and members of the European and Japanese space agencies.

Neurophysiology and Behavioral Physiology

- The etiology of space motion sickness should be identified.
  - Changes in vestibular, otolith, and labyrinth function in microgravity should be characterized.
  - Changes in task performance should be correlated with changes in vestibular and otolith function in microgravity.
- Drug development and testing to prevent or ameliorate the untoward effects of space travel, such as space adaptation syndrome or bone demineralization, should be made a high priority.
- Other possible effects of space flight on neurosensory and biobehavioral function are unknown and should be explored if we intend to achieve a permanent human presence in space.

Bone, Endocrine, and Muscle Physiology

- Changes in the neurohumoral responses to microgravity should be characterized and correlated with the incidence of space motion sickness or changes in task performance.
- The relationships between skeletal muscle atrophy and bone demineralization should be explored using bed-rest and inflight studies.

Hematology

- Erythropoietic, lymphocytic, and granulocytic changes associated with microgravity should be characterized.

- Functional changes in immunology and susceptibility to infectious diseases should be correlated with any qualitative or quantitative changes in hematopoietic cell lines.

Logistic and Policy Strategies

- NASA should give biomedical research the highest priority in its preparations for future missions, particularly for manned missions of long duration.

- NASA should have an active role in the Federal Coordinating Committee in Science and Technology.

- Better integration should be achieved between NASA biomedical research programs and the physical science programs; this integration should relate to spacecraft and Space Station design, as well as to planning of specific experiments in biomedicine.

- The numbers of flights and flight crew personnel available for biomedical research should be increased.

- A national laboratory in space should be established as part of the Space Station and any lunar or Mars base; this laboratory should have designated, well-equipped facilities available to make the full range of measurements required for clinical research.

- NASA should provide better publicity for its biomedical programs. Consideration should be given, for example, to annual meetings cosponsored by NASA and the National Institutes of Health on such topics as "Man on Mars" or "Man in a Space Station." Such efforts should be well publicized in the extramural scientific community.

- Consideration should be given to developing a program involving Specialized Center of Research (SCOR) units in space medicine.

  — This approach should be aimed at developing a number of centers that could be funded for 5 years on a renewing basis similar to the SCOR program at the National Institutes of Health, with a total dollar cost of at least $10-$15 million/year.

  — Such centers should concentrate on multidisciplinary efforts and work that can proceed regardless of delays in flight opportunities.

- Closer ties should be fostered between biomedical researchers at NASA and a broad range of extramural biomedical scientists.

  — Consideration should be given to expanded peer-review committees and external advisory panels and a more formalized and better publicized extramural grants program.

  — Given the extent of biomedical research conducted by foreign space agencies (16), extramural scientists should be encouraged to work with members of the scientific community of the European Space Agency, the National Space Development Agency of Japan, and the Soviet Space Agency, as well as NASA, and NASA should facilitate these interactions.

- Formal linkages should be established between the biomedical research programs at NASA and other agencies, particularly the National Institutes of Health.

- An interagency Space Medicine Coordinating Committee should be developed that would include biomedical scientists from NASA, the United States Air Force Space Command, and other organizations with mutual interests in space research.

- NASA should consider the establishment of NASA professorships for junior and senior university faculty appointees, and these professorships should be supported by the Agency. Such professorships might be referred to as "NASA Professor of Physiology" (or "Microgravity Physiology") in XYZ University. Similarly, NASA should consider the development of awards for faculty training or established investigators similar to the faculty development programs of the National Institutes of Health or the American Heart Association.

Reference List

1. National Academy of Sciences. National Research Council. Space Science Board. 1970. *Life Sciences in Space.* Ed. H. Bentley Glass. Washington, DC: National Academy of Sciences.

2. National Aeronautics and Space Administration. Life Sciences Division. Space Medicine and Biological Research Branches. September 1984. *Life Sciences: A Strategy for the 80's.* Washington, DC: National Aeronautics and Space Administration.

3. Pitts, John A. 1985. *The Human Factor: Biomedicine in the Manned Space Program to 1980.* NASA SP-4213. Washington, DC: National Aeronautics and Space Administration.

4. Proceedings of the Annual Meeting of the IUPS Commission on Gravitational Physiology. *The Physiologist* 27(1984), 28(1985), and 30(1987).

5. Physiologic Adaptation of Man in Space: VII International Man in Space Symposium, February 10-13, 1986, Houston, TX. Sponsored by National Aeronautics and Space Administration, Universities Space Research Association, Baylor College of Medicine, and International Academy of Astronautics. Ed. Albert W. Holland. *Aviation, Space, & Environmental Medicine* 58(September 1987).

6. The Spacelab Experience: Life Sciences. *Science* 225(July 13, 1984):205-234.

7. National Academy of Sciences. Committee on Space Biology and Medicine. 1987. *A Strategy for Space Biology and Medical Science for the 1980s and 1990s.* Washington, DC: National Academy Press.

8. Federation of American Societies for Experimental Biology. July 1983. *Research Opportunities in Cardiovascular Deconditioning: Final Report Phase I.* Ed. M.N. Levy and J.M. Talbot. NASA Contractor Report 3707. Washington, DC: National Aeronautics and Space Administration.

9. Federation of American Societies for Experimental Biology. July 1983. *Research Opportunities in Space Motion Sickness: Final Report Phase II.* Ed. J.M. Talbot. NASA Contractor Report 3708. Washington, DC: National Aeronautics and Space Administration.

10. Volpe, Massimo, Alberto Cuocolo, Filippo Vecchione, Alessandro F. Mele, Mario Condorelli, and Bruno Trimarco. 1987. Vagal Mediation of the Effects of Atrial Natriuretic Factor on Blood Pressure and Arterial Baroreflexes in the Rabbit. *Circulation Research* 60:747-755; Lockette, Warren, and Bruce Brennaman. 1988. Atrial Natriuretic Factor, Vascular Permeability, and Space Adaptation Syndrome. Paper read at Annual Scientific Meeting of the Aerospace Medical Association, May 8-12, at Washington National Airport, Washington, DC.

11. Kohl, Randall Lee, ed. 1985. *Proceedings of the Space Adaptation Syndrome Drug Workshop, July 13-14, 1983,* Lunar and Planetary Institute. Houston: Space Biomedical Research Institute, USRA Division of Space Medicine.

12. Connors, Mary M., Albert A. Harrison, and Faren R. Akins. 1985. *Living Aloft: Human Requirements for Extended Spaceflight.* NASA SP-483. Washington, DC: National Aeronautics and Space Administration.

13. Federation of American Societies for Experimental Biology. April 1984. *Final Report Phase III. Research Opportunities in Bone Demineralization.* Ed. S.A. Anderson and S.H. Cohn. NASA Contractor Report 3795. Washington, DC: National Aeronautics and Space Administration.

14. Hollifield, John. 1984. Potassium and Magnesium Abnormalities: Diuretics and Arrhythmias in Hypertension. *American Journal of Medicine* 77 (Supplement 5a, November):28-32.

15. Federation of American Societies for Experimental Biology. April 1984. *Final Report Phase IV. Research Opportunities in Muscle Atrophy.* Ed. G.J. Herbison and J.M. Talbot. NASA Contractor Report 3796. Washington, DC: National Aeronautics and Space Administration.

16. *USSR Space Life Sciences Digest* 1-14 (1985-1987). Trans. Management and Technical Services Company. NASA Contractor Report 3922.

William DeCampli, M.D., Ph.D.
Chairperson

Francis D. Moore, M.D.

Mark H. Phillips, Ph.D.
Staff Associate

Radiation

Long-duration manned space flight and colonization of nearby objects in the solar system will involve the exposure of humans to a number of environmental stresses, one of which is radiation. Such missions will result in the exposure of astronauts to levels and types of radiation not often encountered on Earth. Different mission scenarios involve different radiation hazards, each of which must be evaluated separately. Our accumulated experience concerning radiation in space, as well as knowledge of radiation hazards gathered on Earth, gives us the means to evaluate some of the radiation hazards to be encountered in space and, more importantly, indicates the limits of our knowledge.

The use of radiation in medicine, and the commercial and military uses of nuclear energy, have led to far-reaching attempts in the United States and other countries to understand radiation and its effects on living creatures. This broad interest means that NASA is not alone in searching for the answers to a number of questions. For many of the relevant measurements and theoretical work, NASA draws on the work of others for its answers. There are, however, a number of problems more or less unique to manned space flight, particularly to missions that extend beyond the Earth's magnetic field for prolonged periods of time. The Earth's magnetic field acts as a shield against the radiation emitted from large solar particle events (SPE's) and from a large fraction of galactic cosmic rays. The radiation from these sources is different in magnitude and biological effect from the radiation sources in the low-Earth orbits (LEO's).

The Radiation Study Group has worked to determine answers to a series of questions. What are the critical problems regarding the effects of space radiation on humans? What is known about the problems? What needs to be known? How can answers be found? This report will do the following: 1) briefly review what is known about the radiation environments in space and the resulting biological responses, 2) define the principal radiation hazards of different categories of missions, 3) assess current research in these areas, and 4) make recommendations for the resolution of outstanding problems in the precise determination of radiation environments and their effects on human health.

The information for this report was obtained from published papers and in response to a solicitation by the Study Group of the views and recommendations

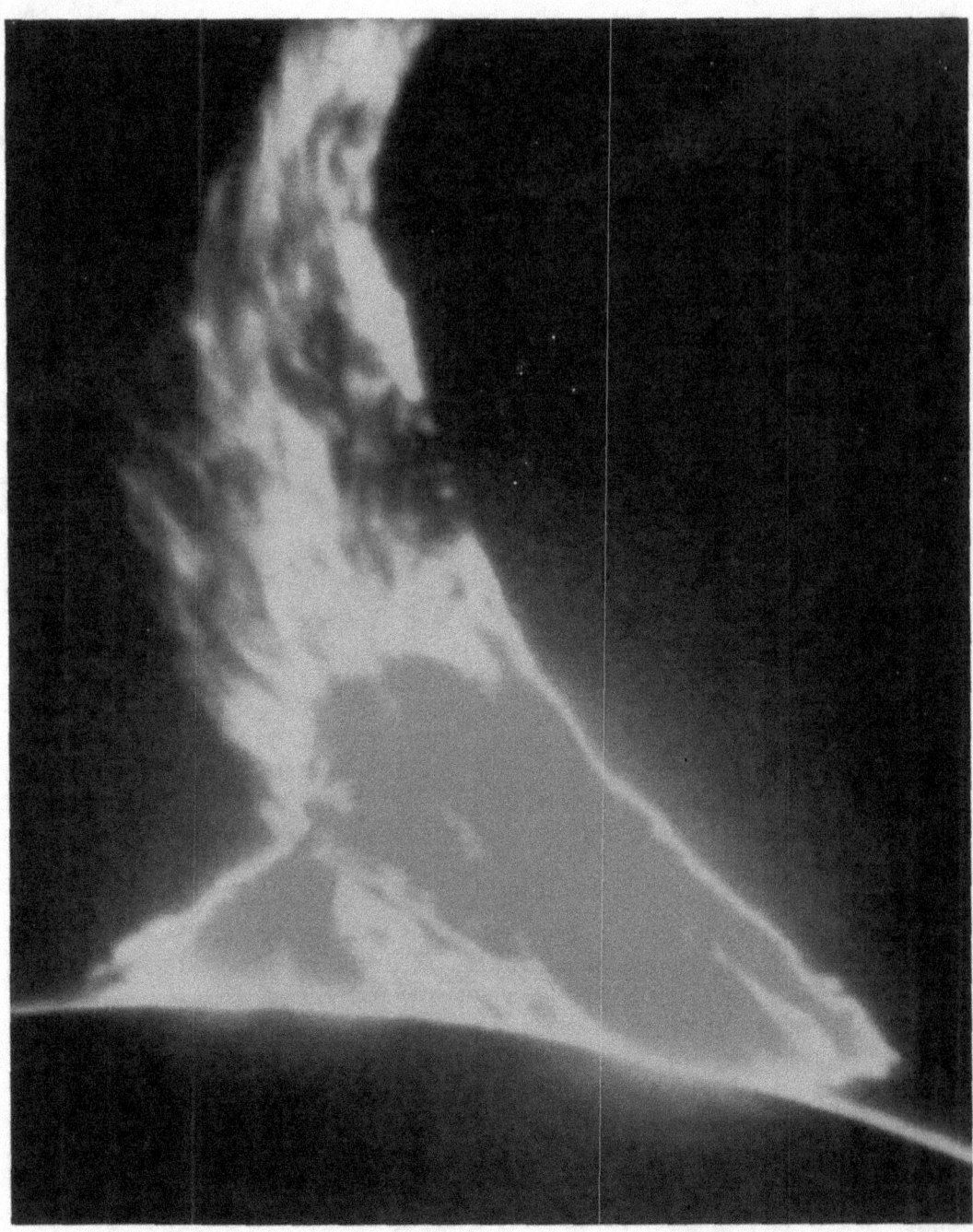

A solar flare erupts on the surface of the Sun. Such eruptions billow forth clouds of particles and other emissions of varying intensity. The Earth's magnetic field acts as a shield for particles emitted by solar events, but space travelers need protection from these emissions and other forms of radiation in space.

of workers in the field. Responses were received from more than 20 prominent scientists at nearly the same number of institutions, including Brookhaven National Laboratory, Columbia University, Goddard Space Flight Center, Jet Propulsion Laboratory, Johnson Space Center, Langley Space Center, Lawrence Berkeley Laboratory, Lawrence Livermore Laboratory, NASA Headquarters,

National Council of Radiation Protection, Naval Research Laboratory, National Oceanic and Atmospheric Administration, Oak Ridge National Laboratory, University of California, University of San Francisco, and U.S. Air Force School of Aerospace Medicine.

Radiation Environments and Biological Effects

The first important problem in determining the radiation hazards to humans in space is defining the radiation environments. A variety of measuring devices have been carried on satellites and manned spacecraft, so that today much is known about the radiation fields encountered in space. Since the fields are dynamic and spatially varying, it is difficult to characterize them completely by measurements. Parallel efforts in modeling are being carried out to provide more complete estimates of these fields.

The space radiation environment is divided into several different categories, depending on the type of radiation and its location. The radiation in LEO is primarily protons trapped in the Earth's magnetic field. In geosynchronous Earth orbit (GEO), trapped electrons and bremsstrahlung radiation produced by spacecraft shielding are the predominant sources. The radiation in both of these sets of orbits varies as a function of position and time. Outside the Earth's magnetic field, radiation comes from large solar particle events and galactic cosmic rays (GCR). Radiation from SPE's occurs sporadically and can be life threatening in intensity. GCR radiation is a low-level, constant background radiation source. The interaction of all of these radiation sources with the material of the spacecraft and its contents alters intensity, spectral characteristics, and quality of the radiation. Table 1 summarizes the sources of radiation.

The effects of radiation on humans are commonly grouped into two categories: acute and long-term. Acute effects include radiation sickness and death; long-term effects are carcinogenesis, teratogenesis, formation of cataracts, and damage to nondividing cells. Figure 1A indicates the types of physiological responses grouped under acute effects and the radiation doses that typically cause their onset (3). Figure 1B describes the temporal pattern of radiation effects following exposure to radiation (3). For long-term missions and colonization, radiation injury to embryos must also be considered. The extent and kind of biological effects depends on the type of radiation, the dose, and the dose rate. In particular, the presence of low and high LET radiation in the space environment has a great impact on the biological effects. The deposition of energy within the cells of the organism is different for low and high LET radiation, resulting in different biological effects. For a given absorbed dose, the relative biological effectiveness (RBE) is a function of radiation type (e.g., photons, particle species) and energy. The RBE also depends on the particular tissue absorbing the radiation.

Mission Scenarios

In the next several decades, a number of different mission scenarios are plausible. As discussed in the preceding sections, the radiation environment unique to each scenario determines the type and magnitude of biological effects to be expected.

Table 1. *Sources of Radiation*

Type	Description	Location	Secondary Radiation
Trapped Electrons	Low LET(1) Large temporal variations Not very penetrating Low dose rate	3-12 E(r)(2) I(max) 3.5E(r) GEO(3)	Bremsstrahlung Penetrating Low LET
Trapped Protons	Ranges from low to high LET Penetrating Low dose rate	LEO(4)	Neutrons High LET
Solar Particle Events	Mostly protons Lesser amounts of heavier ions Occurs sporadically Occasional events of extremely high intensity	Outside Earth's magnetic field— polar orbit GEO Moon interplanetary space	Neutrons Nuclear fragments High LET Penetrating
Galactic Cosmic Rays	Protons, He, heavier ions (especially Fe) Very penetrating, high LET Low dose rate, isotropic	Outside Earth's magnetic field— polar orbit GEO Moon interplanetary space	Neutrons Nuclear fragments High LET Penetrating

(1) LET: linear energy transfer—a measure of the amount of energy deposited as radiation interacts with matter; for a given radiation dose, biological effects are strongly dependent on the LET of the radiation
(2) E(r): Earth radius, equal to 6,000 km
(3) GEO: Geosynchronous Earth orbit
(4) LEO: Low-Earth orbit

The purpose of this section is to define the principal radiation hazards as part of the prelude to outlining what needs to be known concerning these environments.

Low-Earth Orbit

One of the principal missions designed for low-Earth orbit is the Space Station. At the projected orbit parameters (450 kilometers, 28 degrees inclination), the main source of radiation will be the trapped protons in the South Atlantic Anomaly, with a much smaller fraction coming from GCR. Data measured with thermoluminescent dosimeters from the Skylab missions (flown at approximately the same altitude but at larger inclination, 50 degrees) indicate that the average daily dose rate is in the range of 60-70 millirads per day (1). At the greater orbital inclination, the dose due to the South Atlantic Anomaly decreases somewhat, and the GCR dose increases due to less geomagnetic shielding. Calculations by S.B. Curtis, et al., for the proposed mission parameters yield doses of 97 millirem per day to the blood-forming organs behind shielding of 1 g/cm**2 Al (2). At these dose rates, long missions (180 days and more) would require careful personal dosimetry to maintain accepted radiation health limits. As the inclination of the orbit increases, geomagnetic shielding decreases and exposure to solar particle radiation and GCR increases.

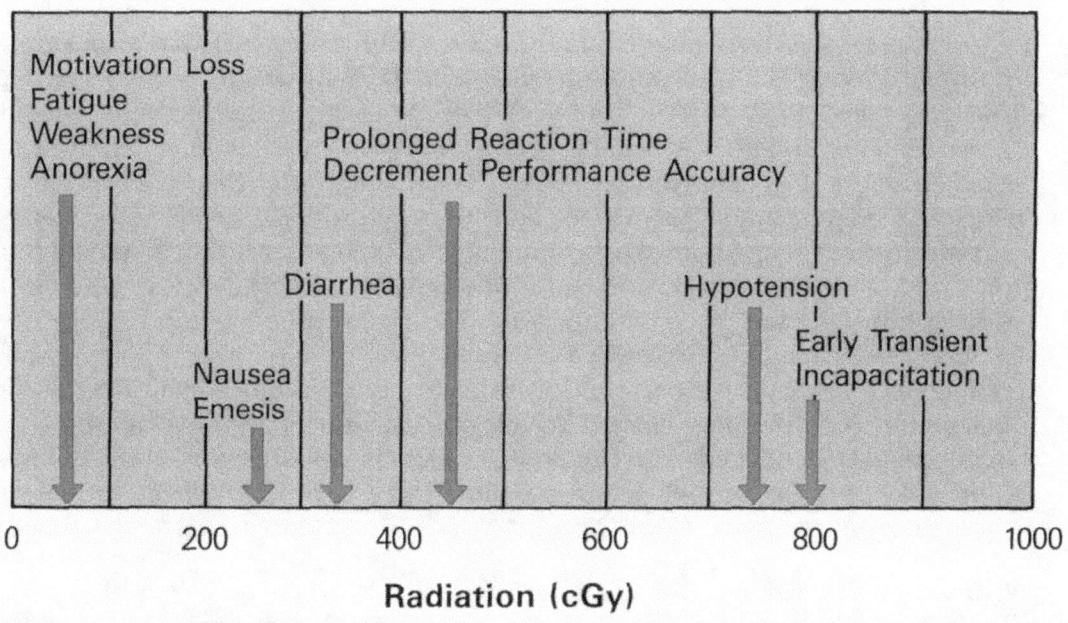

Figure 1A. *Performance Degradation Events Occurring at Various Doses of Radiation*

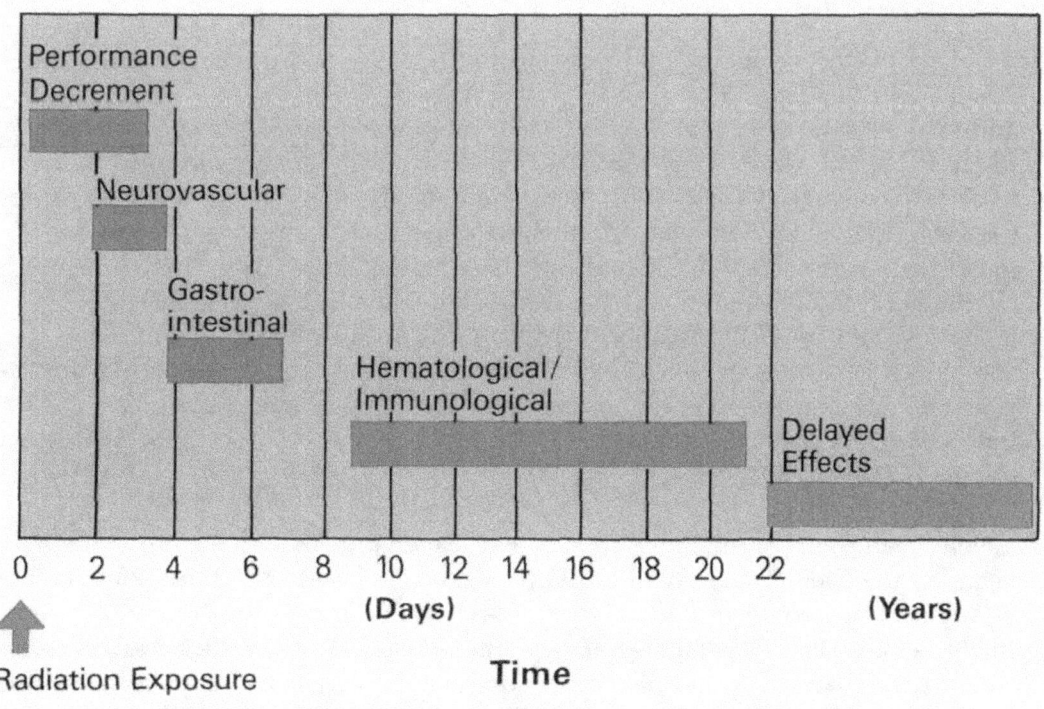

Figure 1B. *Temporal Relation of Postexposure Medical Effects*

Geosynchronous Orbit

A mission in geosynchronous orbit around the Earth faces radiation from several sources: 1) electrons in the outer radiation belt, 2) bremsstrahlung from electron-shielding interactions, 3) GCR, and 4) SPE's. The electrons are high energy, and doses rise very rapidly as shielding decreases to values of less than 2 g/cm**2 (at 1 g/cm**2, the dose is approximately 5 rads/day). With greater than 2 g/cm**2 of shielding, bremsstrahlung dominates and not much is gained with additional shielding; doses range from tens to hundreds of millirads per day, depending on the parking longitude (4). Compared to the first two sources, GCR contributes a smaller but, nevertheless, significant dose. Because of the contribution of fast particles, shielding does not have a profound effect on dose rate. A rough estimate of the dose rate is on the order of 100 millirem per day with no shielding and 50 millirem per day behind 4 g/cm**2 Al (5). (The corresponding physical doses are approximately 30 millirads per day and 20 millirads per day, respectively.) Doses from SPE's vary considerably, corresponding to the wide range of magnitudes of the events.

Curtis tabulates a number of doses for solar particle events occurring in Solar Cycle 19 (1958 - 1961). For shielding of 2 g/cm**2 Al, skin doses were typically 100-200 rads, and doses 4 cm deep in tissues were in the range of 20-50 rads. Behind 5 g/cm**2 Al, the corresponding doses were 20-80 rads and 15-30 rads, respectively. These doses are of a magnitude sufficient to produce acute effects.

Finally, doses from all sources received in transit from LEO to GEO are somewhat less than 1 rem (6).

Lunar Colony

A colony on the surface of the Moon would have no natural magnetic or atmospheric shielding from galactic cosmic rays or SPE's. At solar minimum, the annual dose-equivalent rate due to GCR is approximately 30 rem per year (7). As discussed above, doses from SPE's can be substantially greater. Because of the penetrating nature of GCR, substantial amounts of shielding are needed to stop the HZE (high atomic number, Z, and high energy, E) component. Nuclear interactions between the GCR and the shielding result in production of neutrons, complicating the dosimetry and the calculation of biological effects. Figure 2 illustrates the dose-depth relationships for GCR in lunar material, indicating the complexity of calculating shielding (7). The cosmic rays are significantly attenuated after tens of grams per square centimeter of shielding. However, nuclear interactions result in the buildup of a significant quantity of neutrons, which have a high biological effectiveness.

Given lifetime exposure limits, it becomes clear that if individuals are to spend years on the Moon, substantial shielding would be necessary. Since some amount of surface time would presumably be necessary to perform the colony tasks, it might be advisable to build sleeping quarters deep below the surface. Very-well-shielded safe havens would also be needed for the occasional giant solar particle event.

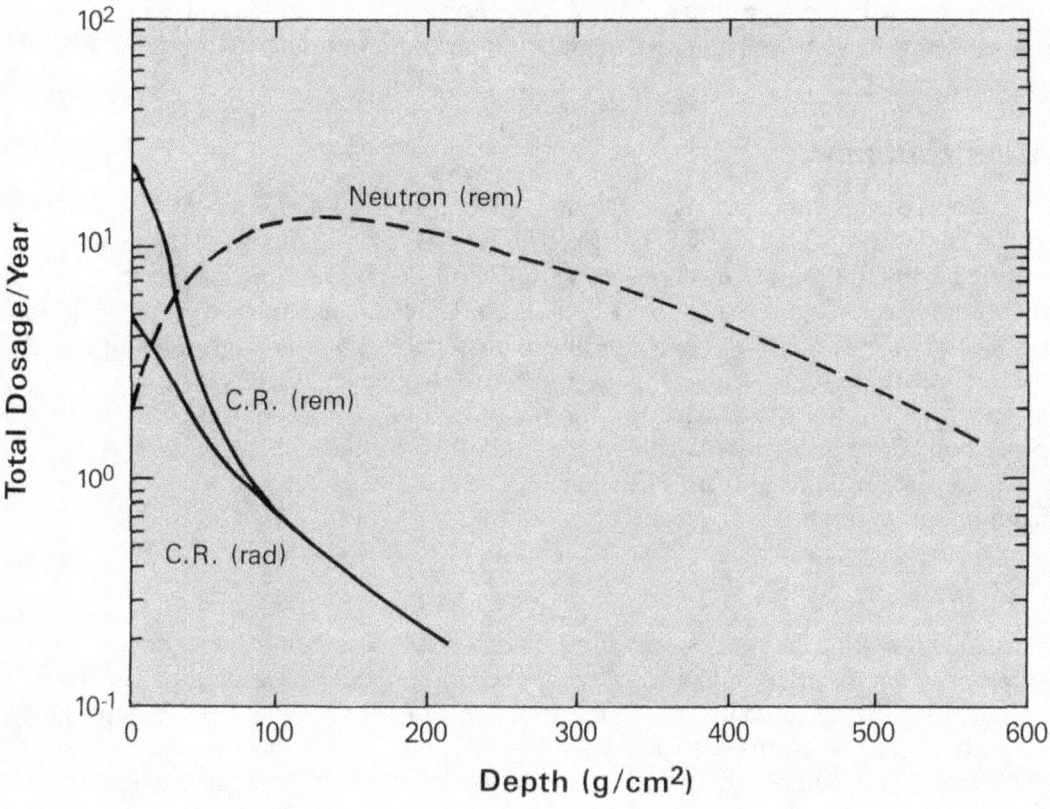

Figure 2. *A comparison of the annual dose equivalent due to secondary neutrons and cosmic-ray nuclei, as a function of shielding; also the absorbed dose rate due to cosmic-ray nuclei*

Mars Mission

The radiation space environment for a mission to Mars is essentially the same as that for a lunar colony, with the exception that during the long space flight, there is no massive shielding readily available in case of a giant SPE. Thus, one must consider the radiation sources as galactic cosmic rays and solar particle events. During the flight, one is limited to spacecraft shielding, and the radiation is isotropic (i.e., there is no shielding of half the solid angle by the planet or Moon). A baseline dose for the Mars trip is 43 rem per year in essentially free space, 36 rem per year behind 4 g/cm**2 Al, and 24 rem per year at the center of a 30 cm diameter sphere of water (8). On its surface, Mars shields half the GCR and the carbon dioxide atmosphere provides some shielding, so estimates of the dose at the surface are approximately 10 rem per year.

As with the other missions beyond the Earth's magnetosphere, the possibility of catastrophic SPE's must be taken into account. More than the other missions, the trip to Mars is especially vulnerable. The trip itself will take on the order of a year to complete, and during that time, there will be no possibility of moving to a lower orbit or burrowing underground. Spacecraft must be designed with a safe haven (a small, very-well-shielded region where the crew may temporarily seek shelter).

Mars Colony

Radiation risks common to those of the previous two missions would be present in establishing a Mars colony. The trip to Mars with its attendant exposures, coupled with the long-term exposure to GCR on the surface of the planet, constitute the radiation exposure. At this point, more consideration must be given to the fact that people may be spending considerable parts of their lives in the colony and that childbearing is likely to occur, given the difficulty of a return trip to Earth. As on the Moon, deep shelters can be built for sleep and SPE protection. But the long-term carcinogenic effect and the cumulative central nervous system damage from HZE particles, mutagenesis, and teratogenesis grow in importance in such a scenario.

Lifetimes in Space

Scenarios resulting in people spending lifetimes in space bring together the radiation risks discussed in practically all the prior sections. Depending on the exact nature of the missions, acute effects from SPE's may be the most important, or perhaps the problems associated with procreation in a radiation environment of HZE particles may be dominant. While the possibility of spending a lifetime in space is at the remote edge of current thinking, the previous scenarios naturally lead to this consideration.

Current Research

As indicated by the preceding sections, a considerable amount of information is available about the space radiation environment. Progress has been made in determining space radiation fields and in modeling the interaction of radiation with shielding, as well as in the radiobiology of both high and low LET radiation.

Radiation Source Determination

The space environment presents four basic categories of radiation, as mentioned above: trapped protons, trapped electrons, SPE's, and GCR. A complete characterization of each type implies knowledge of the spatial distribution, particle fluence, spectral distributions of energy, variations in fluence (particles/area/time) and spectrum with time, and (for SPE's and GCR) the relative amounts of different ion species.

Free space radiation interacts with the spacecraft and shielding materials. Measurements are made within satellites or spacecraft, and it is often difficult to account for all the varying amounts of shielding surrounding the dosimeters. Once estimates of the shielding are made, modification of the radiation field as it passes through the shielding must be accounted for in determining the free space environment.

Detector response affects the accuracy with which the radiation field is measured. Each kind of detector is limited by the type of radiation it can measure and the amount of detail concerning the categories of radiation it can measure. As our knowledge of the space radiation environment has grown, so have appropriate detectors been designed, built, and improved. Thermoluminescent detectors have been developed that measure low LET radiation well, but not high LET. Plastic track detectors have been devised to identify the HZE component of space radiation. Experience in nuclear physics is resulting in the construction of detectors capable of determining the atomic number and energy of charged particles in radiation fields.

When the energy from solar particle events reaches the Earth, it can cause geomagnetic storms that disrupt broadcast communications and can result in aurora such as the one photographed by astronaut Robert F. Overmyer during the Spacelab 3 mission.

Although it is difficult to map radiation fields, much is now known about the radiation surrounding Earth. Trapped electrons, occupying a much wider range of altitudes—some extending many Earth radii away—are perhaps not quite so well measured as the protons. The pronounced interaction of electrons with shielding material resulting in the production of bremsstrahlung also complicates these measurements. Fairly solid data exist for the spatial distribution of electrons as a function of altitude and their spectral distributions at each altitude. It is known that the particle fluence undergoes marked diurnal fluctuations, as well as strong variations influenced by solar storms. This discussion perhaps can best be summarized by the observation that current space radiation data and models can predict the radiation measurements on the Space Shuttle only within a factor of two (9).

Measurement of SPE's is probably the most uncertain aspect in determining the space radiation environment. The frequency of their occurrence is related to the solar cycle, so that accurate characterization is to some extent determined by the length of the solar cycle (11 years). As the name implies, this type of radiation is not continuous but occurs in short bursts of several days. The timing of their occurrence is of much interest to the humans in space program, but as yet, the events cannot be predicted. The temporal evolution, spectral characteristics, and particle species profile are all subject to variations and are functions of a number of variables. So far, the ability to predict the magnitude of the events is very limited. In an effort to establish an early warning system for astronauts, however, work is progressing on correlating measurements of the electromagnetic emissions and the time course of the initial parts of the event to the eventual absolute magnitude of the event. Finally, efforts to determine accurately all aspects of SPE's are made difficult by the effect of the Earth's magnetic field on the protons and heavier ions emitted.

Characterization of the galactic cosmic rays shares many of the same problems as the SPE's, namely the effect of the Earth's magnetic field and the need to determine both the spectral distribution and the ion species profile. However, GCR is isotropic and continuous, though the magnitude is affected by the solar cycle. Because HZE particles are not common on Earth, instrumentation adequate to measure them was not developed as early as for the lower LET components of space radiation.

Shielding

The interaction of electrons with matter and the production of bremsstrahlung have been well understood as a result of work in the medical field. The transport of protons and HZE particles in matter has been studied only since the development of particle accelerators. Early accelerators produced just protons and helium nuclei at the energies of interest in space research, and it was not until the 1970's that an accelerator (the Bevalac at Lawrence Berkeley Laboratory) was developed capable of producing heavier particles at energies similar to those found in GCR. Research in these areas has been led primarily by nuclear and atomic physicists. More recently, NASA's experience in space and the medical applications of charged particle beams have spurred work in charged particle transport.

The work in this area has proceeded along both theoretical and experimental lines. Measurements at accelerators and reactors, as well as in space, have yielded much information on radiation interactions in matter. Theoretically, research is progressing on combining the measured data with physics theory into models capable of predicting the type, magnitude, and distribution of energy deposition, the nuclear interactions between incoming radiation and target nuclei that produce secondary radiation, and the spatial distribution of both primary and secondary radiation. Radiation transport computer codes are currently used in evaluating mission design parameters and evaluating radiation risks. They reflect, of course, the uncertainties in the measured radiation fields mentioned above, as well as in the theoretical aspects. The transport of HZE particles is particularly uncertain, owing to the relative newness of the field and the lack of data for many ion species at a range of energies.

Biological Effects

The effects of radiation on humans is the research area fraught with the most uncertainties. This is the result of a number of factors, such as the difficulty of defining endpoints in complex organisms, the fact that humans cannot be used in prospective experiments, and the lack of experience with the quality of radiation and low dose rates encountered in space.

Research is conducted on a variety of levels: biomolecular, cellular, tissue systems, animal models, and humans. The impetus for this research lies in the need to understand basic interactions between living beings and radiation, to exploit radiation effects for medical ends, and to understand the risks associated with medical, industrial, and military uses of radiation. Hence, a wide range of organizations is involved in radiation research.

The biological effects of low LET radiations are much better understood than those of high LET radiations. Many data exist on radiation effects on the suborganism level and animal models, but it is not always easy to extrapolate those results to humans. One of the major effects of low-level radiation is carcinogenesis. Much work has been done in this area, but results are often difficult to interpret because of the high natural incidence of cancer and the presence of numerous confounding factors. Recent work indicates that the effects of exposure to low dose rates of high LET radiation are quite different from the effects of low LET radiation. This may have a profound impact on space missions, and extensive research needs to be done.

The establishment of RBE's for different radiations and different tissues is currently the subject of a number of experiments, but the task is far from complete. The importance of microlesions induced by high LET radiation is another subject that needs to be understood more fully.

Findings and Recommendations

Solar Particle Events

Findings

- For all of the proposed missions, except the Space Station in LEO, the possibility exists of the mission crew being exposed to debilitating or lethal doses of radiation as a result of solar particle events.

- The degree of our ignorance of these events, coupled with the potentially disastrous consequences to both the crew and the mission, establish SPE's as the most pressing challenge for the humans in space program.

- Much work needs to be done to characterize fully the flux, spectral distribution, and time evolution of SPE's. In addition, support should be available for astrophysical studies and solar modeling work relevant to establishing an early warning and prediction system.

Recommendation

- **NASA should vigorously pursue basic research in solar physics in order to model and predict catastrophic radiation events and to investigate short-time warning systems that will provide time for the crew to seek protection.**

Radiation Biology

Findings

- Much needs to be learned about the radiobiological effects of high LET radiation, an issue central to establishing a long-term human presence in space.

- The importance of this research to NASA stems from a number of factors: the pervasiveness of GCR and secondary radiation in space environments relevant to NASA missions, the high biological effectiveness of high LET radiation, the differences in effects between low and high LET radiations, and the early stage of this field's development.

- Work is needed to establish the relative biological effectiveness for HZE particles, to investigate the low dose-rate effect of high LET radiation relevant to such topics as carcinogenesis, cataractogenesis, embryonic development, and the functioning of the nervous system, and to provide a basic theoretical basis for radiobiology and track structure.

- Additional attention needs to be directed to the development and evaluation of radioprotectors, the interaction between ionizing and ultraviolet radiation, and possible interactions between environmental stresses to the organism and radiation.

Recommendation

- **NASA should vigorously pursue basic research in the radiation biology of high LET radiation.**

Shielding and Transport

Findings

- More complete knowledge of radiation-shielding interactions is necessary to determine radiation risk factors for the mission crew and to design adequate protection.

- This effort requires measurement of the free-space radiation environment, measurement of the radiation environment within the spacecraft, and accelerator-based experiments designed to study the interaction of radiation and matter.

- Parallel research efforts in the modeling of these interactions will result in transport codes (computer programs that simulate the passage of each type of radiation through defined series of materials) that can be used for the design and evaluation of a range of situations.

Recommendation

- **NASA should direct the following efforts to work in shielding and transport research: conduct measurements of the free-space radiation environments; study the interaction of radiation with shielding materials through the development of the transport computer codes and accelerator experiments. A balanced approach in studying the free-space radiation** environment, the radiation environment inside the spacecraft, and accelerator-based experiments is desirable.

Instrumentation and Measurement of the Space Radiation Environment

Findings

- Improvements are needed in both passive dosimeters (devices that measure cumulative exposure and are processed at intervals) and real-time dosimeters (devices that provide automated and continuous measurements of radiation). The development of appropriate biological dosimeters (a system that measures a change in a biological endpoint) is also an important priority.

- An effort needs to be made to measure accurately the free-space radiation environment, so that uncertainties in measurements behind shielding can be removed and the data can be applied to arbitrary shielding situations.

- The space radiation environment beyond the Earth's geomagnetic shielding needs to be characterized further, as does the electron flux in GEO.

Recommendation

- **NASA needs to support basic research in instrumentation and measurement of the space radiation environment.**

Research Support

Findings

- NASA's interest in radiobiological issues has become focused over the years, and it is clear that there are some overriding issues in which the Agency has considerable stake.

- NASA has no focused program on the biological effects of radiation, but there are unresolved issues in this field critical to the success of the Agency's current and future missions.

Recommendations

- **NASA should make a commitment to support fundamental research on the biological effects of radiation. This support and commitment should take the form of expanding NASA's role in and funding for basic research and of contributing to the necessary facilities, such as the Bevalac accelerator.**

- **NASA should continue to function as focal point for the wide range of radiobiological research activities relevant to its needs. To maintain its leadership role, the Agency should encourage collaborative efforts with other organizations and agencies interested in similar areas of research, including the Department of Defense, the National Institutes of Health, and the National Oceanic and Atmospheric Administration.**

Reference List

1. Benton, R.P., and R.P. Henke. 1983. Radiation Exposures During Space Flight and Their Measurement. *Adv. Space Res.* 3:171.

2. Curtis, S.B., et al. In press. Radiation Environments and Absorbed Dose Estimation on Manned Space Missions. *Adv. Space Res.*

3. Conklin, J.J., and R.I. Walker. 1987. Military Radiobiology: A Perspective. In *Military Radiobiology*, ed. J.J. Conklin and R.I. Walker. Orlando, FL: Academic Press.

4. Stassinopoulos, E.G. 1980. The Geostationary Radiation Environment. *Journal of Spacecraft and Rockets.* 17:145.

5. Silberberg, R., C.H. Tsao, and J.H. Adams. 1984. Radiation Doses and LET Distributions of Cosmic Rays. *Rad. Res.* 98:209.

6. Curtis, S.B. 1974. Radiation Physics and Evaluation of Current Hazards. In *Space Radiation Biology and Related Topics*, ed. C.A. Tobias and P. Todd.

7. Silberberg, R., et al. 1985. Radiation Transport of Cosmic Ray Nuclei in Lunar Material and Radiation Doses. In *Lunar Bases and Space Activities of the 21st Century*, ed. W.W. Mendell. Houston: Lunar and Planetary Institute.

8. Letaw, J.R., R. Silberberg, and C.H. Tsao. 1986. Natural Radiation Hazards on the Manned Mars Mission. In *Manned Mars Missions, Working Group Papers*, NASA M002.

9. Schimmerling, W. Telephone conversation with Mark Phillips, March 1987.

William C. Schneider, D.Sci.
Chairperson
Gerald P. Carr, P.E., D.Sci.
Michael Collins
Peter B. Dews, M.D.
Lauren Leveton, Ph.D.
Staff Associate

Crew Factors

Within the next decade, NASA will plan enterprises that place small groups of humans in space for extended periods of time. The success of extended missions will require a thorough knowledge of how to establish conditions that enhance human capabilities for living and working in space for prolonged periods of isolation and confinement. This paper examines the major issues associated with crew factors, particularly those issues associated with long-duration space flight: crew/environment interactions, interpersonal interactions, human/machine integration, crew selection, command and control structure, and crew motivation.

Several assumptions are made about the characteristics of groups assigned to long-duration missions in space. Crew size will most likely be small, with fewer than 10 crew members. Mission lengths will vary, but Space Station crew rotations of 60 to 180 days are being proposed, while a Mars mission will require isolation and confinement for a 1- to 3-year period. In addition, many of the missions under consideration — Mars, a lunar base, and even the Space Station — entail only a limited possibility of emergency rescue and return to Earth.

For the crew, long-duration space flight, such as on the Space Station and future missions, will require separation from customary physical and social environments and confinement within a highly limited and sharply demarcated environment (1). This isolation and confinement, which is experienced in some similar ways by submarine crews and Antarctic field research teams, produce stress, which can increase as the mission lengthens. The stress, in turn, can result in boredom, depression, irritability, increased anxiety, disturbed sleep, fatigue, hostility, and lowered motivation (2,3,4). These symptoms reduce crew effectiveness and productivity.

Problems are now being recognized about the ways in which available space capsules and systems affect human capabilities to perform effectively within a small, confined, and isolated group in extended microgravity conditions. Preliminary reports from long-duration Soviet missions are disquieting. It is significant that concerns are being expressed by senior NASA administrators who are not themselves life scientists. The expressed concern is no longer only about physiological survivability, but also about the environment and systems needed for humans to work effectively as part of a group and to maintain psychological health under the projected conditions of space flight (5,6).

The solutions to these problems are as much beyond our direct experience as were the effects of even an hour of microgravity before the space program began. Let us assume that the physiological problems associated with 1 or more years spent in microgravity or low gravity are survivable or that the outstanding engineering problems, as specified in the "Systems Engineering" paper that follows, have been resolved. The major challenge to behavioral scientists is this: how to design and program the hardware, environment, and activities to keep the crew productive, psychologically healthy, and satisfied. Crew selection and training will continue to remain important. It is probable, however, that the maintenance of positive, productive relationships among the crew will become a more important issue during extended missions.

Current plans to extend the presence of humans in space have highlighted limitations in the knowledge about the psychological, social, and behavioral requirements for successful long-duration manned missions. This section identifies major scientific issues in the area of crew factors.

Crew/Environment Interactions

Systematic research has been limited concerning how best to organize teams, tasks, and the environment to enhance crew efficiency and satisfaction.

The challenging features of life in space for astronauts on extended missions include the danger and risk of the mission, the constancy of the environment, prolonged confinement and isolation in close quarters, the similarity of the daily schedule, the lack of privacy, and the limited number of constant companions. These circumstances place demands on selecting, training, and organizing the crew and in engineering the environment (7,8,9). In missions conducted to date, great care has been taken in the first three areas. The crew members have been assigned to and trained for particular tasks in a well-organized unit led by a commander. The environment has been determined by the exigencies of the work stations

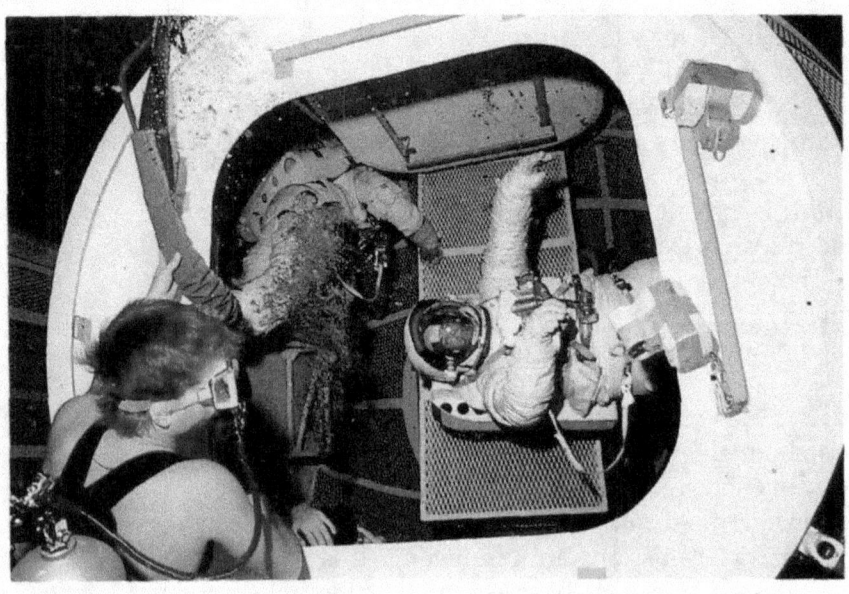

The Weightless Environment Training Facility is used for crew training in a simulated space environment.

and attempts to make the spacecraft as habitable as feasible, largely in response to the wishes of astronauts and space human factors engineers (10,11,12).

Among the areas associated with human space flight, the two most amenable to modification are the design of the spacecraft interior for work and leisure and the scheduling of activities. While the former has received increasing attention (10), the latter has been left largely to mission requirements. Neither has been subjected to systematic ground-based studies with full crews involved in realistic simulations of actual missions. The importance of proper scheduling in enhancing crew cooperation and performance must not be underestimated. An important aspect of work scheduling is determining the best mixes of automated and manual control, discussed later in this paper.

Interpersonal Interactions

The major issue is identifying the requirements needed to maintain psychological health, sustain relationships, and optimize performance among the crew during long-duration missions.

A considerable amount of information is available concerning the effects of confinement and isolation on the performance, cohesion, and well-being of small groups. There is difficulty, however, in generalizing the results from laboratory studies of small groups to crews that will be living in spacecraft for extended periods of time. Current studies cannot assess the danger associated with long-duration missions, as compared to the safe conditions of the scientific laboratory. Many important questions remain that are pertinent to interpersonal interactions in confined, isolated, and high risk environments.

The most significant of these questions involves identifying the ways to sustain cooperative and satisfying interactions among crew members throughout an extended mission. Among other factors affecting group relationships, such as the age, sex, and education of crew members, we do not have sufficient information to predict with confidence the optimal size for a group to travel to Mars or to establish a lunar base (3,4). In addition to group size, role definition is important. Clearly defined roles consistent with the statuses of group members are important in achieving optimal performance. The command and control structure of the space crew will play an important part in the group's role definition (13). It is probable that the conditions of space travel will require creative solutions best developed by group members with diverse backgrounds and capabilities.

Research examining interaction patterns among different types and structures of groups needs to be conducted. Groups must be studied living in conditions and performing activities that approximate as closely as possible the environment of the spacecraft and workload of a space mission. Actual performance variables, as well as interactive variables, must be examined. Ground-based studies on a large scale over a long period will be needed to obtain baseline, normative data. It is imperative that these efforts begin at once.

Human/Machine Integration

One of the major issues in designing human/machine systems is determining the requirements for space systems in which the crew can work effectively.

Many compromises in manned spacecraft designs have lowered human productivity. A first step, then, is to determine designs for effective performance. Environments for living and working in space must then be developed that will help sustain crew performance throughout long-duration space missions.

One important area is the relationship between human and automated tasks during extended missions. At present, well-established principles do not exist to guide the distribution of tasks between human and automated systems for maximum efficiency and reliability. What is the effect on crew productivity and morale of an increasing dependency on machines to perform tasks and make decisions? Will this become more of a problem in a long-duration space flight? Can unforeseen combinations of inputs to an automated system lead to seriously inappropriate outputs? How can such eventualities be safely aborted by human interventions? Uninformed assignment of tasks to the crew and automated systems may compound problems caused by human fallibility and automated inflexibility.

An understanding of the human/machine interface and its effects on productivity also involves recognition of group and individual performance factors. The effects of human error may be exaggerated by increasingly complex, automated systems. A trend toward more complex and autonomous missions with fewer human operators may make the remaining human tasks all the more taxing. To enhance crew safety, the potential for human error and automated inflexibility has to be fully understood and controlled.

The extensive ground-based research on design of work stations and the selecting and training of users must continue to be incorporated into the development of the Space Station and spacecraft for long-duration missions. Specific developments recognizing special requirements associated with microgravity need to be the subject of intensified efforts. Astronauts and former astronauts with experience in space should be involved in guiding the research.

Development of human performance models, through anthropometric and biomechanic design considerations, can provide information about body dimensions and mobility important in reducing or preventing human error. In addition, effective user selection and training can help reduce errors by matching the characteristics of the user as closely as possible to system design characteristics. A major problem regarding education in complex autonomous systems is that the human cannot be trained in detail for everything. Therefore, the orientation must be less specific and involve some system accounting and tolerance of error.

Crew Selection

Criteria for selecting crew members for long-duration, self-sufficient space missions need to be addressed.

One of the most important issues in planning long-duration missions involves developing criteria for selecting space crews. The criteria need to make possible the identification of personnel who will perform well in a group setting over a long period of time. Final validation of the criteria will come from assessment of the outcome. Researchers must continue to study groups in isolation to assess and develop predictors of performance as well as work efficiency. It will be necessary to study ground-based groups in isolated, confined, and potentially high risk environments and in other conditions simulating as accurately as possible future missions.

Part of the selection process involves screening crew members for specific positions, particularly those of commander and second-in-command. Choosing commanders for long-duration space flight requires reevaluation of current selection procedures (7,8). Coming up through the ranks is no longer the only appropriate strategy. Mission leaders will have to be chosen on their ability to create and manage the conditions for optimal crew performance during extended space missions, involving prolonged periods of confinement and isolation (7,8). Research should be conducted to define the qualities requisite for positions of crew leadership.

Important training issues and related concerns are also outstanding. Among the questions are the following: What kinds of training should individual crew members have in small group behavior? Should NASA provide a psychological support team to help monitor and maintain the well-being of the crew, as do the Soviets? How can we ensure that crews are compatible through selection and training processes? In addition, how can the environment be engineered to sustain cooperative behavior? Specifically, how can crew tasks, schedules, and programs be designed to maintain cooperation among the crew? Another issue, among still others requiring research and resolution, concerns the contingency plans needed if communications among members of the group break down and the mission becomes jeopardized.

Command and Control Structure

An effective command and control structure for ensuring success in long-duration missions needs to be identified.

The major issue in this area involves the authority structure in the spacecraft during the mission (13). An initial concern is the commander of the space crew. As suggested above, the attributes and skills that would qualify individuals as effective leaders for extended missions are unlikely to be the same as current commander attributes. In addition, the appropriate guidelines for exercising authority during interpersonal conflicts among the crew that threaten the mission's success need to be defined. A related issue involves the following question: Will a command structure, either within a space crew or between ground command and the space crew, survive the full year or two of isolation before Earth systems can directly affect the crew? The problem is compounded by the close quarters within the space vehicle. Moreover, long delays in communication will be characteristic of a Mars mission. Will such a situation cause a shift in locus-of-control and, if so, when will it be most likely to occur? How can such a shift be managed? Other questions relate to the phases of the mission (i.e., transit to Mars, onsite effort, return to Earth) that may require a different partitioning of authority.

These issues can no more be resolved in terms of current, direct experience than could questions concerning the effects of prolonged microgravity in earlier times. While this lack of information will not deter individuals from volunteering for missions, attempts must not be neglected to discover ways to reduce risks. Research should be conducted concerning various ways of organizing space crews. Validation will be forthcoming when long-term flights are conducted.

Crew Motivation

The major issue concerns how best to enhance human productivity through environmental design solutions and optimal scheduling of tasks.

Maintaining high levels of motivation and performance among group members presents special problems in the stressful and confined environment of space. The effects of long-term isolation and confinement can be significant. However, the acknowledgment of such effects has been notably missing in the official reports of American and Soviet space flight experiences. The information that is available is anecdotal. It has been speculated that astronauts are reluctant to acknowledge instances of decreased performance, and space program officials are disinclined to acknowledge behavioral problems publicly (13).

Informal reviews of mission reports and interviews with space crews and ground personnel provide the outlines of the larger picture. While overall performance has been remarkably good, decrements have been evidenced in experimental errors, lost data, equipment mishandling, and a variety of behavioral disturbances, including sleep loss, fatigue, irritability, depression, anxiety, mood fluctuation, boredom, social withdrawal, motivational shifts, and fatigue-induced crew conflicts (10,12,13,14,15,16,17).

A number of important questions relate to motivation and performance. What, for example, are the motivational factors that influence human performance in long-duration missions? Other questions include the following: What kinds of work, rest, and recreation schedules are needed to keep the crew occupied, motivated, and satisfied? How can mission planners ensure that crew members will continue to perform effectively as a team? What kinds of training, task scheduling, and selection criteria will provide effective countermeasures to problems in crew coordination? What is the best strategy for attaining and maintaining optimal crew motivation and performance?

The work, rest, and recreation schedule is a crucial factor in minimizing performance degradation. While performance progressively deteriorates as a function of flight length and rest never completely restores performance during a long-duration mission, a proper work/rest schedule can minimize and retard this process (18,19).

Time management and the sequencing and arranging of interactions and activities are fundamental to crew compatibility and motivation. An important aspect of this is crew workload. Workload problems have been evident throughout the manned space program (12). Overload leads to dissatisfaction and to decreased performance, which in turn can compromise and endanger a mission. Underload, or too little work, also causes difficulties, for it can affect crew morale negatively and waste valuable opportunities. Important questions include: What workloads are required for extravehicular activity (EVA) operations? Are the requirements too demanding? What kinds of tasks and activities can be designed to keep crew members active and highly motivated during long-duration missions?

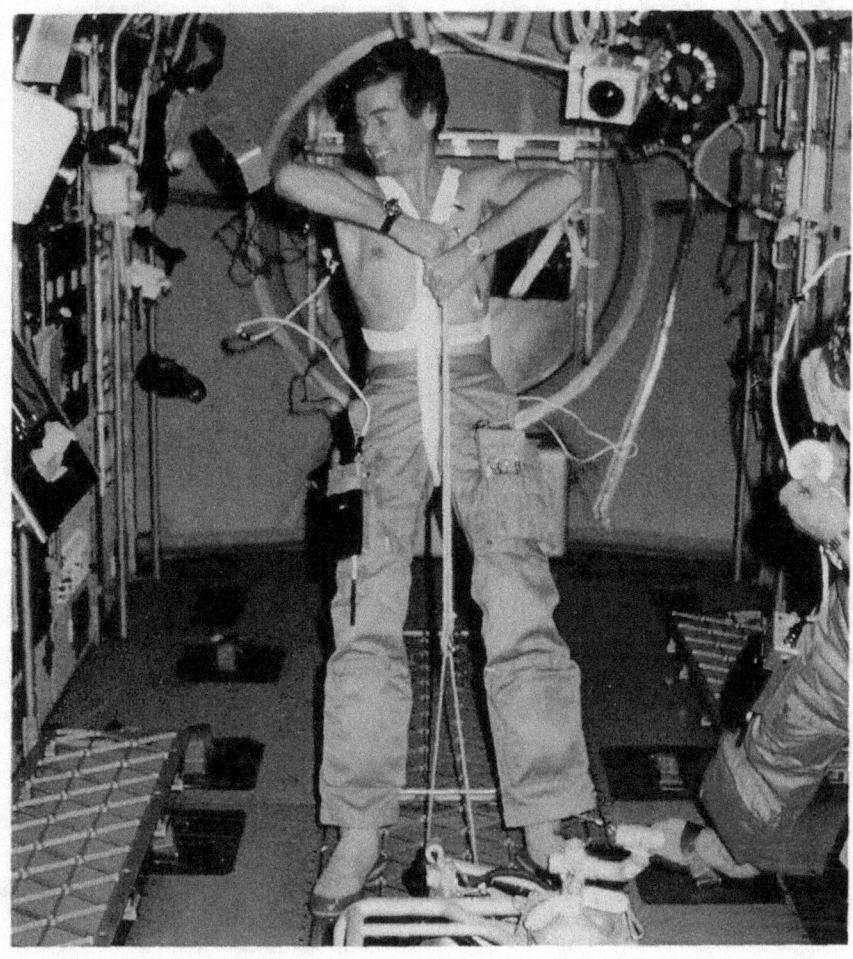

Payload Specialist Ulf Merbold works out with the Spacelab 1 exercise facility.

As noted earlier, the crew will spend more time monitoring increasingly automated and complex systems, which can result in boredom and frustration and in a significant performance problem. Measures should be developed to provide relief from highly monotonous and routine tasks. Task design and assignment should be studied carefully to avoid problems of workloads that are too demanding or overly monotonous. In addition, research is needed to develop interactive work programs between the crew and the scientific and technical apparatus of the mission. The activities must approximate real tasks and be skillfully programmed so that the interactions between operator and machines are a key factor in sustaining performance. Research is also required in crew fatigue, particularly the relationship between fatigue and decreased performance and the types and scheduling of tasks to circumvent problems associated with fatigue.

Findings and Recommendations

Crew/Environment Interactions

Finding

- The problems associated with motivating space crews and maintaining their efficiency and satisfaction will probably increase as missions become more lengthy. Very little information exists on how environmental configurations and the programming of activities can enhance crew productivity and morale.

Recommendation

- NASA should continue research into the influences of environmental configurations and the programming of activities on crew efficiency and morale.
  - The research should involve ground-based simulations of space mission modeling and other analog situations.
  - Particular attention should be paid to determining optimum combinations of automated and manually performed mission tasks.

Interpersonal Interactions

Findings

- Small group interaction on extended space missions is an important issue.
- The Space Station, as well as ground-based analogs, can provide an opportunity to collect information about the dynamics of space crews that can be applied to future long-duration missions.

Recommendations

- Research should be based on existing data and information from Soviet and American space flights, undersea habitats, submarines, Antarctic expeditions, and other analogous settings. Additional research should be performed in laboratory and field settings.
- Studies should be made concerning the effectiveness of confined and isolated groups that vary in size and composition, especially according to male/female ratios, ethnic diversity, and the education and skills of members.
  - The groups should be studied in conditions (i.e., physical, temporal, and social) that approximate the spacecraft environment.
  - The dynamics of crews on the Space Station should be studied to gain information that can be applied to future long-duration missions.

Human/Machine Integration

Findings

- Well-established and validated principles do not exist currently for effectively distributing tasks between human and automated systems.

- Future space-flight missions will involve the use of more complex and autonomous systems with fewer human operators. The potential for human error may be increased.

Recommendations

- Crew/environment interactions should be studied intensively to provide the basis for designing space systems that will elicit and sustain optimal crew performance.

- Information needs to be obtained from rigorous scientific study of crew members in prolonged space-flight conditions and in analog research settings that will help determine the factors related to optimal crew performance. Direct access to crew members is required to assess the factors that influence crew performance and psychology. The information resulting from such efforts is essential to designing living and working environments that will maximize crew productivity.

- Studies that examine the allocation of functions between humans and machines to enhance crew performance during space flight, particularly during long-duration missions, should be continued. Operating systems should be designed to accommodate human error.

Crew Selection

Findings

- It is necessary to understand how to select members for small groups that must work and live together for prolonged periods in isolated and confined environments.

- The selection process needs to include group training in team building and crew coordination, communication skills, and crisis management.

Recommendations

- Current investigations, space flight, and analog settings information should be the basis for intensive, directed research.

- Small groups should be studied in increasingly realistic situations for longer times to identify predictors and training that make for group success.

Command and Control Structure

Finding

- The success of long-duration missions will depend in part on the effectiveness of the crew's authority structure.

Recommendation
- Research should continue on the effects of different command structures on crew performance and psychological health and the relationships between ground command and the space crew, particularly regarding possible shifts in locus-of-control patterns.

Crew Motivation

Finding
- It is necessary to develop the means to maintain high levels of crew motivation throughout long-duration space missions.

Recommendations
- Research should be intensified on the variables that influence individual and crew productivity. These variables include the causes of performance decrements, such as certain types and amounts of work, task scheduling, and crew fatigue.
- An assessment should be made of work requirements and task scheduling to achieve and maintain a high level of crew motivation and performance.
- Empirical research is needed to determine the most effective work/rest schedules for extended-duration missions on the Space Station and other future space-flight missions.

Reference List

1. Regal, D.M. 1986. Human Performance in Space. In *Proceedings of the Human Factors Society 30th Annual Meeting,* 1, 365-369. Santa Monica, CA: Human Factors Society.

2. Bluth, B.J. 1981. Soviet Space Stress. *Science 81* 2:30-35.

3. Bluth, B.J. 1982. The Psychology and Safety of Weightlessness. Paper presented at the 15th Symposium on Space Rescue and Safety, Paris, France.

4. Bluth, B.J. January 1986. Sociology on the Space Station. *Space World* 8-10.

5. Alexander, Joseph K., Philip C. Johnson, Percival D. McCormack, David C. Nagel, Sam L. Pool, M. Rhea Seddon, Joseph C. Sharp, and Frank M. Sulzman. January 1987. *Advanced Missions with Humans in Space.* No city of publication given: National Aeronautics and Space Administration.

6. National Aeronautics and Space Administration. Office of Space Science and Applications. Life Sciences Division. A.E. Nicogossian. October 1984. *Human Capabilities in Space.* NASA Technical Memorandum 87360. Washington, DC: National Aeronautics and Space Administration.

7. Chidester, T.R., and H.C. Foushee. 1987. Selection for Optimal Performance in Aerospace Environments. In *Abstracts of Space Life Sciences Symposium: Three Decades of Life Sciences Research in Space,* 234-235. Washington, DC: National Aeronautics and Space Administration.

8. Ginnett, R.C. 1987. Is "The Right Stuff" Right?: The Leader's Role in Crew Formation and Development. In *Abstracts of Space Life Sciences Symposium: Three Decades of Life Sciences Research in Space,* 235-237. Washington, DC: National Aeronautics and Space Administration.

9. Hackman, J.R. 1987. Group and Organizational Influences on Crew Effectiveness. In *Abstracts of Space Life Sciences Symposium: Three Decades of Life Sciences Research in Space,* 237-239. Washington, DC: National Aeronautics and Space Administration.

10. Clearwater, Y.A. 1987. Human Factors Design of Habitable Space Facilities. Paper presented at 38th Congress of the International Aeronautical Federation, October 10-17, Brighton, United Kingdom.

11. Stuster, J.W. 1986. *Space Station Habitability Recommendations Based on a Systematic Comparative Analysis of Analogous Conditions.* Report No. 3943. Washington, DC: National Aeronautics and Space Administration.

12. Wise, J.A. 1986. The Space Station: Human Factors and Habitability. *Human Factors Society Bulletin* 29:1-3.

13. National Academy of Sciences. Committee on Space Biology and Medicine. 1987. *A Strategy for Space Biology and Medical Science for the 1980s and 1990s.* Washington, DC: National Academy Press.

14. National Academy of Sciences. National Research Council. Committee on Human Factors. 1983. *Research Needs for Human Factors.* Washington, DC: National Academy Press.

15. McNeal, S.R., and B.J. Bluth. 1981. Influential Factors of Negative Effects in the Isolated and Confined Environment. Paper presented at the Fifth Princeton/AIAA/SSI Conference on Space Manufacturing, May.

16. Bozhko, Andrey. March 1987. Special Feature: Group Dynamics/Psychology Diary of Participant in Soviet Isolation Experiment. *USSR Space Life Sciences Digest* Issue 10:91-98. Trans. Lydia Hooke. NASA Contractor Report 3922(12).

17. Connors, Mary M., Albert A. Harrison, and Faren R. Akins. 1985. *Living Aloft: Human Requirements for Extended Spaceflight.* NASA SP-483. Washington, DC: National Aeronautics and Space Administration.

18. Alluisi, E.A. 1986. *Research in Performance Assessment and Enhancement.* Report No. ITR-69-12. Arlington, VA: Army Behavioral Science Research Laboratory.

19. Ray, J.T., O.E. Martin, and E.A. Alluisi. 1960. *Human Performance as a Function of the Work/Rest Cycle: A Review of Selected Studies.* NRC Pub. No. 882. Washington, DC: National Academy Press.

William C. Schneider, D.Sci.
Chairperson

Gerald P. Carr, P.E., D.Sci.

Michael Collins

Peter B. Dews, M.D.

Jay P. Sanford, M.D.

Lauren Leveton, Ph.D.
Staff Associate

Systems Engineering

This paper addresses the critical life sciences aspects of systems engineering. As used in this paper, systems engineering is the art and science of designing environments, systems, facilities, and products to support the health, safety, performance, and productivity of crews. The involved activities, as well as related life sciences efforts that are also part of systems engineering, become increasingly important as missions are planned that extend the time humans spend in space and their independence from ground-supplied resources.

NASA's Program in Systems Engineering

Systems-engineering activities related to life sciences are dispersed throughout NASA's organization. They range from basic research, to applied science, to technology development. The relevant activities in basic research and applied science are primarily organized under the Life Sciences Division's Space Medicine and Biology Program. In addition, some activities are conducted in the Advanced Technology Development Program. Issues related to the extended presence of humans in space are receiving increasing attention from NASA Headquarters organizations, particularly the Office of Aeronautics and Space Technology, the Office of Space Station, and the Office of Space Flight. Ames Research Center provides most of the basic research and technology advancement. Johnson Space Center, in coordination with Marshall Space Flight Center, conducts the more applied research activities and operations.

Scientific Issues

The Systems Engineering Study Group had a wide range of disciplines within its purview. Given limitations in time and resources, it concentrated on four areas representing key engineering concerns related to the life sciences: Crew Protection and Health Systems, Extravehicular Activity (EVA) Systems, Habitability Requirements, and Space Adaptation/Gravity Environment. The previous discussion, "Crew Factors," reviewed other systems-engineering issues, including the human/machine interface.

Crew Protection and Health Systems

Crew protection and health systems include the environmental-monitoring and

Astronaut Sherwood C. Spring, positioned on the end of the remote manipulator arm, checks joints on the assembly concept for construction of an erectable space structure tower. The tower in the photograph extends from the cargo bay of the Space Shuttle Atlantis.

decontamination, radiation protection, and life support technologies required to maintain a safe and healthful environment within the spacecraft.

Environmental-Monitoring and Decontamination Systems. The major purpose of these systems is to monitor, detect, and prevent any contamination problems within the spacecraft environment that could threaten the health and safety of the crew. Spacecraft materials behavior during long-term habitation, water treatment chemicals, materials processing, and biological and experimental activities increase the probability that contaminants will be released into the closed environment that may ultimately threaten crew health (1).

Routine monitoring of air and water quality, particularly for trace contaminants, as well as the microbial environment, will be needed beginning with the Space Station. Except for gas composition, however, NASA does not have the requisite technology available. A recent assessment of environmental-monitoring and control requirements has identified deficiencies in the following areas:

- Buildup of microbial flora on the spacecraft surface and EVA systems
- Environmental debris in terms of volatile organic compounds, airborne particulate matter, and metals
- Microorganisms and the buildup of treatment chemicals and leached contaminants in recycled waste water
- Fire within the spacecraft
- Vibroacoustics control.

Real-time monitoring systems, particularly sensors, are required to detect and characterize contamination levels from these factors. All the potential environmental hazards need to be clearly identified and the means for effectively counteracting them need to be developed. Acceptability standards also should be determined. In addition, procedures for maintaining a nontoxic environment need to be developed, and the crew should be trained in implementing these procedures.

NASA should decide if it needs to develop the environmental-monitoring and decontamination technology itself. The risks involved with using existing technologies on the Space Station need to be clearly evaluated, particularly since it is known that the current instrumentation is marginal. It is critical to understand fully how the environmental quality requirements will change as missions to the Moon and Mars are planned. In addition, it is important to investigate the contingencies required and to establish the responsibilities for managing the needed actions in the event of severe contamination in the spacecraft environment. Cleaning materials constitute another potential source of hazard. A study should be conducted to determine if there are synergistic effects that will be detrimental to crew health.

Other significant issues to be resolved in maintaining environmental quality include the impact of monitoring tasks on crew performance. Will the available instrument technology require too much of the crew's time? A related issue involves the effective allocation of monitoring tasks between humans and machines. How can we build systems to compensate for complacency errors?

Radiation Protection. Extended missions involving humans in space are permissible only if the crew is protected from unacceptable exposure to ionizing radiation, as is indicated in the "Radiation" section of this report. Central concerns within systems engineering are understanding protective requirements and developing effective environmental design solutions for preventing exposure to ionizing and non-ionizing radiation within the spacecraft and during EVA operations (2). A protective system needs to be devised that will shelter the crew from radiation, particularly during periods of high flux, and still allow members to accomplish required tasks. The entire spacecraft cannot be designed for worst case flux levels because of unacceptable weight and volume penalties. Part of the spacecraft, however, might be designed to provide the shielding necessary for missions lasting to and beyond 1 year, should the Nation decide to embark on such ventures.

Life Support Systems. Mission duration is the most significant factor determining the type of life support systems required on spacecraft. To date, NASA's manned missions have been short enough for life support functions to run on consumable supplies. Of these supplies on manned missions, water and air account for the greatest volume and mass. Although first generation technology exists to partially recycle water and regenerate air, these supplies and the food needed to sustain crews are carried on the spacecraft or, for permanent missions in low-Earth orbit, they can be resupplied from Earth. Regenerative life support systems could be used on the Space Station to reduce logistic requirements and operating costs. Development costs would be significant, however. Nevertheless, some form of bioregenerative or "closed-loop" system must be used for long-duration missions, such as a lunar or Martian colony, as discussed in the "Controlled Ecological Life Support Systems" (CELSS) section of this report.

One of the key engineering issues is integrating the life support system within the spacecraft and developing the capability to isolate the system from any contamination problems. The integration and isolation requirements must be developed early in the design process.

We have learned that the Earth's ecosystem has considerable resiliency and tolerance for abuse. Because of its relatively small size and limited variety of life forms, the closed environment in a space vehicle is vastly different. System resiliency is restricted, and the margins for design error and performance variation may be extremely small. Consequently, research and acquisition of experience in closed cycle, environmental life support systems is one of the most important requirements confronting space life sciences. Until we can build and depend on a life support system that will tolerate dynamic interaction with a human crew, we cannot embark upon extended missions to the Moon or to other planets. The type of partially closed life support system envisioned for the Space Station cannot meet the requirements of a lunar base or a Mars mission. Therefore, it is important to implement a research and technology effort to develop options for closed, regenerative life support systems.

EVA Systems

The work needed to develop EVA systems that can be maintained in orbit is challenging and must be expedited. Suits will have to be maintained on the Space Station, a requirement that cannot be met using present EVA systems. The particular type of system to be developed will be determined according to such parameters as the kinds and amounts of work needed. The more there is to do, the higher the premium on efficient operations. Currently, 6 hours of EVA satellite maintenance have been factored into the guidelines (3). However, EVA operations may be much more rigorous than these guidelines allow. The endurance of astronauts during servicing activities sets the upper limit on how much EVA can be accomplished. One factor limiting astronaut endurance has been suit design. In soft suit technology, higher pressures result in decreased flexibility, particularly in the hands. In hard suit technology, joint mobility solutions appear promising, but dexterity problems remain unresolved.

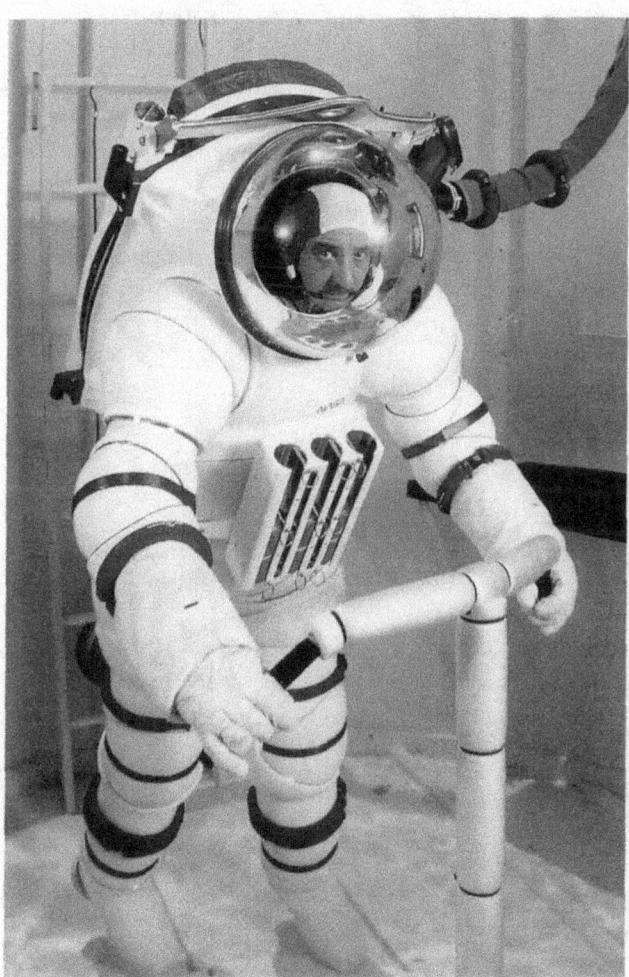
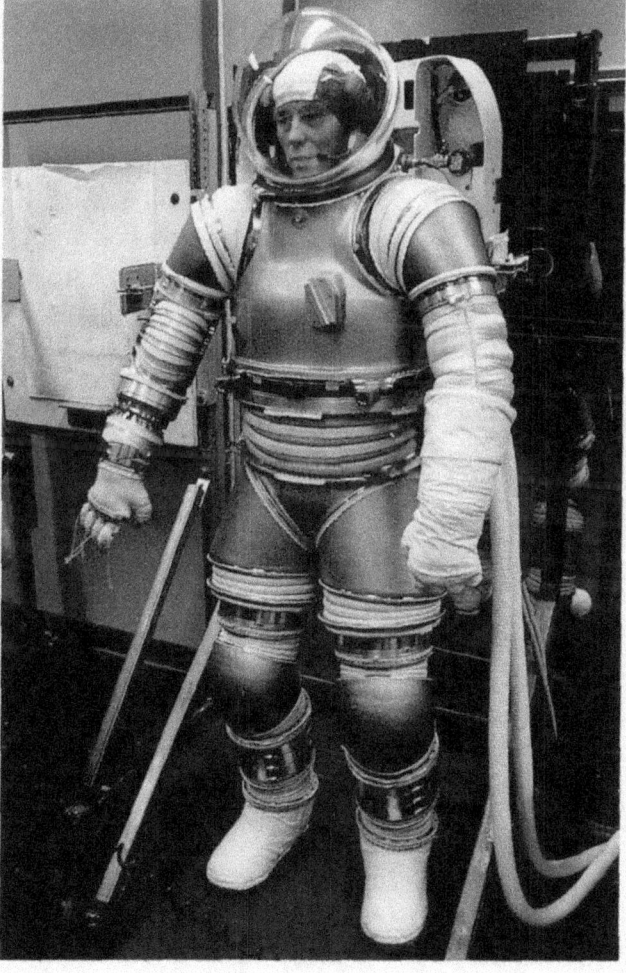

These photographs illustrate two prototypes of next-generation spacesuits. Vic Vykukal tries out a suit designed at ARC that employs hard-suit technology. At right, Astronaut Jerry Ross wears a hybrid soft/hard suit designed at JSC. The relatively high pressures within these suits allow astronauts to conduct extravehicular activity without lengthy decompressurization from the atmosphere of a spacecraft.

The design of EVA systems should begin with an analysis of the requirements for conducting the EVA tasks. Recognizing the problems and accepting the risks involved in EVA operations is critical to such an approach. The need is for a high capability system, which has the potential to encourage growth in satellite-servicing operations and other EVA activities associated with the Space Station. Use has been made of an anthropomorphic suit for zero-gravity activity. An EVA enclosure concept, sometimes called a "man in a can," may, however, be more effective for most situations. An atmospheric pressure room inside the enclosure would allow for a variety of behaviors (e.g., eating, resting, scratching), and use of an integrated locomotion system could greatly reduce physical exertion and lengthen EVA time. For tasks requiring dexterity, a number of end effectors must be developed; prehensors and gloves should be tailored to the jobs to be done.

EVA systems for surface use will present different challenges. They must allow mobility in 1/6 or 1/3 gravity, withstand the wear from surface dust and chemicals over 1 to several months, withstand and function in a high CO_2 atmosphere (Mars), and be light enough to be worn and carried by an astronaut under the prevailing gravity conditions for full work days. The weight of the portable life support system (backpack) must be addressed. Provisions must also be made for backpack regeneration and suit servicing on the surface of another planet.

At present, no one group within NASA is evaluating the entire question of EVA. A comprehensive look at the direction of EVA operations is clearly needed to identify the requirements of future activities and to develop EVA systems capable of satisfying these requirements.

Habitability Requirements

Habitability involves the design of environments to support and enhance crew productivity, performance, health, safety, and comfort (4,5). Early research in habitability focused on such factors in space flight as temperature and humidity, sensory deprivation, and variable acceleration (6). Current studies of spacecraft habitability emphasize the relationship between technological and human factors (7). The extent to which environments are congruent with the needs and preferences of the individual determines the degree of person-environment fit, or habitability. The following list identifies the major spacecraft factors pertinent to habitability and the well-being of the crew:

- Volume
- Temperature and humidity
- Lighting
- Vibroacoustics
- Personal hygiene and waste management
- Privacy
- Aesthetics or functional decors
- Food systems
- Leisure and recreation
- Environmental monitoring and control.

All of the above factors, and other human factors design requirements, are discussed in the four volumes of *Man-Systems Integration Standards (MSIS)*, issued by NASA in March 1987 as NASA-STD 3000. This document, which is a significant first step in developing a standardized set of human factors requirements, will be revised as necessary to include additional standards for future manned space activities. It is important to recognize that the *MSIS* is a compilation of what seems to have worked in the past. As such, it is based largely on experience. The document needs, however, to be enhanced by systematic testing of alternatives to determine the optimum, as is done for the more purely engineering specifications of the spacecraft. In addition, it is vital that instrumentation be available to measure all key aspects of the actual environment so that proper control can be exercised.

"Crew Factors," the previous discussion, explored the psychological and sociological ramifications of long-duration space flight. A major issue for systems engineers is how to design the environment to enhance the psychological health of the crew. Prolonged periods of confinement and isolation are psychologically damaging if deficiencies exist in the livability, or habitability, of the environment. For example, inappropriate noise and vibration levels, inadequate water and food systems, privacy constraints, recreation activities incompatible with crew preferences, and an aesthetically monotonous environment can have a profound influence on the psychological health of individuals in confined and isolated settings (7,8,9). These factors represent potential sources of stress that can lead to low morale, decrements in performance, and an increased vulnerability to illness.

Food may pose additional problems during space flight. It becomes an increasingly important concern, on psychological, physiological, and technological bases, as mission duration lengthens. Nutrition is an important factor in maintaining physiological health. Manipulations of the crew's diet may even be an effective countermeasure for some of the degenerative effects of weightlessness. Beginning with the Space Station, food must be stored for increasing lengths of time, utilizing methods that do not require much space or electric power and that minimize system weight and size. There is a logistic penalty for any significant amount of food that is not consumed.

The food preparation system has to be flexible enough to allow for a variety of alternatives and self-selection and to require minimal preparation time either by one person or the entire crew for individual, special dinners, emergency rations, and group meals. For the Space Station, the best system may be to store food in bulk and prepare meals from ingredients. The important point is that Space Station technology in this area has yet to be developed, as is the case with other areas of habitability, including the hygiene and waste management systems. While much work is currently under way on the food preparation system, the efforts must be expanded. Food and food preparation will be a vital factor in the success of any long-term space mission.

The Space Station represents an opportunity to validate and extend our understanding of the relationships among habitability factors systematically so that control over them for future long-duration missions will be possible. To accomplish this, new methods are needed to obtain the significant habitability assessment data. Some of the important questions include the following: Who will have access to the crew and their environment? How often will access be possible and what types of data can be collected? How will the habitability of the Space Station be evaluated? The Space Station can be used as a scientific laboratory to answer these questions and to define the requirements for interplanetary missions. As many factors as possible should be studied in ground-based simulations to ensure maximum usefulness of the expensive and limited resources available on the Space Station.

Space Adaptation/Gravity Environment

Long-term missions require that crew members safely adapt and readapt to varying gravity conditions. Scientific evidence is lacking at present to demonstrate that the provision of partial gravity may prevent or reduce the effects of microgravity exposure. A variable-gravity research facility is required to support basic investigations of the efficacy of fractional gravity in attenuating the effects of repeated or prolonged exposure to microgravity conditions. The salient questions include the following: What changes are there in crew productivity and performance following prolonged exposure to microgravity conditions? Can a human live comfortably and work productively in a partial-gravity facility that has a fixed- or variable-rotation rate? What is needed to ensure crew comfort? How much artificial gravity is needed and for how long to maintain crew performance and productivity during long-duration missions? What are the major engineering problems in developing a rotating facility? Should a variable-gravity facility be used before the crew descends to the surface of the Moon, or perhaps Mars, or as a recovery vehicle after the flight?

The problems of adaptation to various gravity environments relate to a number of interesting engineering challenges. An important issue is identifying the requirements for making a large, rotating spacecraft a habitable and productive environment. These requirements are not presently known. It is important that NASA establish a research and development program to provide the basis for designing a rotating Mars transit vehicle.

Findings and Recommendations

Environmental-Monitoring and Decontamination Systems

Findings

- The possible contaminants in a spacecraft are many, ranging from toxic gases to particulate matter. The environmental-monitoring system must be able to monitor accurately the status of all critical environmental factors in the same fashion as the Health Maintenance Facility monitors the health of the crew.

- The success of long-duration missions will depend in part on knowing the impact of continual environmental monitoring of crew performance, the successes and limitations of technologies to be used on the Space Station, the kinds of systems required for missions to the Moon and Mars, and the types of countermeasures that will help the crew resist the health hazards associated with pathogenic bacteria that may build up during space flight or other possible contamination problems.

Recommendations

- NASA must support development of an environmental-health-monitoring system capable of detecting all possible sources and types of contamination or other life-threatening factors from air, water, and food systems. Included in the former category are toxic and microbial contaminants. Additional hazards involve problems associated with radiation and fire, vibroacoustics, debris, and thermal regulation systems.

- The research and development program for environmental monitoring and decontamination should investigate countermeasures to help the crew resist the health hazards associated with contaminants and other life-threatening factors. In addition, the program should investigate the contingencies required and establish the responsibilities for managing the needed actions in the event of a contamination problem.

Radiation Protection

Finding

- The variety of radiological hazards, primary and secondary at different locations within a spacecraft, are not known with sufficient precision to make adequate engineering specifications for shielding possible.

Recommendations

- Research should be undertaken to measure the radiation more precisely during missions at various locations both within and outside the spacecraft.

- Studies should be made not only of crew health but also of crew productivity with the use of such radiation protective measures as water tanks.

- NASA should increase support for research into the development of experimental design solutions for limiting the crew's exposure to radiation.

Life Support Systems

Findings

- Closed-loop life support systems (i.e., regenerative systems for air, water, food, and the absorption of carbon dioxide), which will become increasingly important for longer term missions, are far from operational. Many key questions require resolution. Systems need to be redundant, using different components, and optimum combinations need to be developed.

- Ground-based research needs to be conducted to develop life support systems, which should be tested on the Space Station, preferably in a life sciences module.

- In addition, life support requirements for a possible lunar base and a Mars mission need to be identified to determine if parallel or separate developmental efforts are required.

Recommendation

- NASA must intensify its efforts to determine the requirements for regenerative air, water, and food systems that could support long-duration missions, such as a lunar base or a Mars mission. The development of these systems should be scheduled so that they can be tested and used on the Space Station. Self-contained portable life support systems must be developed for use in space missions and on the surfaces of planets.

EVA Systems

Findings

- NASA has a clear requirement for a significant increase in EVA operations in the next 20 years.

- Research into EVA operations is not sufficiently emphasized within NASA and needs to include the best cross section of experts.

Recommendations

- NASA should focus its research and development program on EVA systems on the following:

    — Defining EVA operations for future missions

    — Identifying clear requirements for these missions

    — Delineating innovative options for optimal EVA systems

    — Developing technology for the identified EVA systems.

- NASA should conduct a study to determine EVA requirements for the next 20 years. A panel should then be appointed to identify approaches for meeting these requirements. The panel should comprise well-known researchers in EVA suit design, perhaps including representatives from the undersea diving industry, and a cross section of experts from Ames Research Center, Johnson Space Center, NASA Headquarters, the Office of Aeronautics and Space Technology, and the Office of Space Science and Applications.

Habitability

Finding

- Systematic study has not been made of specific habitability requirements, such as the amount of space required per person to maintain crew productivity and well-being for lengthy missions and the relationship between environmental stress and human tolerance and errors.

Recommendations

- Systematic studies of habitability requirements should be considerably expanded to identify outstanding issues and to provide information applicable to long-duration space flight and the potential success of the Space Station as a habitable vehicle.

  — This process should incorporate available data from the Astronaut Office and from personnel involved in Antarctic expeditions, submarine missions, and Soviet space flight.

  — Additional information should be elicited from ground-based simulations, as well as underseas habitats and polar stations.

- A systematic research program should be established to utilize fully the unique capabilities of the Space Station in delineating human habitability factors for long-duration space missions.

- NASA should allocate funding each year for updating *Man-Systems Integration Standards*. The information in these volumes will be important in meeting the requirements of long-duration missions.

Space Adaptation/Gravity Environment

Findings

- Long-term missions require that crew members adapt and readapt successfully to varying gravity conditions.

- Scientific evidence is lacking at present to demonstrate that the provision of partial gravity may prevent or reduce the effects of exposure to microgravity.

Recommendation

- Research should be conducted to identify the requirements for designing a large, rotating spacecraft that is safe and habitable.

Reference List

1. Alexander, Joseph K., Philip C. Johnson, Percival D. McCormack, David C. Nagel, Sam L. Pool, M. Rhea Seddon, Joseph C. Sharp, and Frank M. Sulzman. January 1987. *Advanced Missions with Humans in Space.* No city of publication given: National Aeronautics and Space Administration.

2. National Aeronautics and Space Administration. March 1987. *Man-Systems Integration Standards.* Vol. 1. NASA-STD-3000. Houston, TX: Johnson Space Center.

3. National Aeronautics and Space Administration. Johnson Space Center. 1987. *Space Station Program: Definition and Requirements.* JSC-3000. Houston: Johnson Space Center; Cohen, M.M., and S. Bussolari. 1987. *EVA Access Facility: A Comparative Analysis of Four Concepts for On-Orbit Space Suit*

Servicing. Vol. 2 of *Human Factors in Space Station Architecture.* NASA Technical Memorandum 86856. Washington, DC: National Aeronautics and Space Administration.

4. Wise, J.A. 1985. *The Quantitative Modeling of Human Spatial Habitability.* NASA Grant No. NAG 2-346. Moffet Field, CA: Space Human Factors Office, NASA Ames Research Center.

5. Wise, J.A. 1986. The Space Station: Human Factors and Habitability. *Human Factors Society Bulletin* 29:1-3.

6. National Academy of Sciences. National Research Council. Space Science Board. 1972. *Human Factors in Long-Duration Spaceflight.* Washington, DC: National Academy of Sciences.

7. Clearwater, Y.A. 1987. Human Factors Design of Habitable Space Facilities. Paper presented at 38th Congress of the International Aeronautical Federation, October 10-17, Brighton, United Kingdom.

8. Connors, Mary M., Albert A. Harrison, and Faren R. Akins. 1985. *Living Aloft: Human Requirements for Extended Spaceflight.* NASA SP-483. Washington, DC: National Aeronautics and Space Administration.

9. Stuster, J.W. 1986. *Space Station Habitability Recommendations Based on a Systematic Comparative Analysis of Analogous Conditions.* Report No. 3943. Washington, DC: National Aeronautics and Space Administration.

Jay P. Sanford, M.D.
Chairperson
Carolyn L. Huntoon, Ph.D.
Ivan L. Bennett, M.D.
Barry J. Linder, M.D.
Staff Associate

Operational Medicine

Operational space medicine focuses on the care of astronauts. Despite limited experience in this area at the beginning of the manned space program, astronaut health care has been successful both during flight and on the ground. As Operational Medicine evolved within NASA's Life Sciences Division, responsibilities increased to include flight and ground health care of astronauts and their families, a longitudinal study of the astronauts' health, a study of spaceflight effects upon the astronauts, as well as development of possible countermeasures to these effects, and identification of the medical aspects of selection and retention criteria for astronauts.

Both the Office of Space Science and Applications (OSSA) and the Office of Space Flight (OSF) at NASA Headquarters have responsibilities for Operational Medicine. Johnson Space Center (JSC) has been delegated the prime responsibility for medical operations, while certain field centers, such as Kennedy Space Center (KSC) and Dryden Flight Research Facility (DFRF), are key support participants within NASA. The Department of Defense as well as several hospitals also are part of the overall support system. In addition, intergovernmental and interagency agreements with the Departments of Commerce and Transportation, the Federal Aviation Administration, the Federal Communications Commission, and other institutions support Operational Medicine. Four NASA Management Issuances and two implementation plans also contribute to program definition. Several advisory committees and boards have a voice as well in defining the structure and policy decisions of Operational Medicine; these bodies include the Life Sciences Advisory Committee, the Committee on Space Biology and Medicine of the National Academy of Sciences (NAS) Space Science Board, the Medicine Policy Board at NASA Headquarters, and the Medicine Board at JSC.

Operational Medicine has worked well in the environment of single missions, most of which have been of relatively short duration, the longest lasting 84 days. An understanding of short-term physiological adaptations to space flight is developing, and appropriate countermeasures are being pursued. However, serious issues concerning the consequences of long-term space flight remain to be answered before Operational Medicine can confidently support humans in space for long-duration missions (over 180 days).

Important assumptions made throughout this paper are that no mission with humans in space can be risk free and that the goal of Operational Medicine must be health risk reduction to a clearly defined level acceptable to the Agency.

The challenge faced by Operational Medicine is to support successfully several simultaneous long-duration missions involving humans. This discussion assumes that NASA will develop the Space Station and proceed eventually to a lunar base and/or a manned Mars mission, perhaps with a new generation of Space Transportation System vehicles (such as Shuttle II and/or the National Aerospace Plane [NASP]). Presently, such missions are generally not limited by technological issues but by a critical lack of data and understanding of the effects of long-duration space flight on humans.

The following sections examine areas important to the continued success of Operational Medicine at NASA. Previous recommendations are reviewed and in many cases endorsed, while new and specific suggestions are advanced. Data for this discussion were accumulated through a review of pertinent documents from NASA Headquarters, NASA field centers, and contractors, as well as from information on medical issues from Soviet space life sciences translations, U.S. submarine experience, and Antarctic expeditions. In addition, interviews were conducted with life sciences officials at NASA Headquarters, JSC, KSC, and Ames Research Center (ARC).

Issues, Opportunities, and Findings

The Operational Medicine Program at NASA has gained practical experience through the successful support of many manned space flights. This paper endorses the current and planned practices of the program through the early Space Station era. Critical issues facing Operational Medicine are mainly concerned with longer duration missions and are detailed below.

Inflight Health Maintenance Facility

Extensive definition and prototype development work is currently in progress for the Space Station Health Maintenance Facility (HMF) (1,2). The facility is more than an emergency room in orbit; its exercise facilities serve a role in

A Health Maintenance Facility such as this prototype will be used on the Space Station to monitor and diagnose crew health.

preventive medicine. In addition, the HMF has capabilities for definitive diagnosis and treatment, such as for minor surgery and dental work. Currently, use of the HMF for clinical biomedical research purposes is under consideration. Accordingly, the dividing line between operational and research usage may become increasingly indistinct, and Operational Medicine must be prepared to deal with this situation if it materializes.

As mission scenarios mature and medical experience increases, periodic review and revision of HMF requirements will be necessary. Extended missions to the Moon or Mars would pose quite different requirements from those of Space Station missions (3). In addition, provision of a Crew Emergency Return Vehicle (CERV) would necessitate reassessment of HMF capabilities.

Future Manned Spacecraft

Since medical requirements for space vehicles can influence engineering design, Operational Medicine needs to contribute to and influence engineering decisions in the early design stages of future manned spacecraft. For example, Operational Medicine concerns for the National Aerospace Plane should be fully addressed (currently, the NASP program has not had any direct communication with NASA Operational Medicine). Additionally, the program will need to define the specific medical requirements for the Orbital Maneuvering Vehicle (OMV) and/or the Orbital Transfer Vehicle (OTV) if they are man-rated.

It is difficult to predict accurately the incidence of medical emergencies on the Space Station, as explained in a status report commissioned by JSC concerning epidemiologic analysis of Space Station disease/event rates (4). Moreover, the Agency has not clearly defined an acceptable level of health risk. It seems likely, however, that if a medical emergency should arise and the Shuttle be unable to arrive in time to effect a successful rescue, the consequences for the Space Station program could be catastrophic.

Two medical emergencies have been recorded requiring use of a return vehicle on a Soviet space station. Given the likelihood of a medical emergency during the life of the U.S. Space Station program, the medical requirements with respect to internal volume, capabilities, reentry profile, and vehicle recovery times of a CERV need to be firmly established. Further consideration must be given to the operational impact of a CERV with additional capabilities, such as one having an ability to function as a safe haven from environmental dangers (including accidental release of atmospheric toxins or pyrolytic products, sudden Space Station decompression, and/or radiation exposure from internal sources or from solar flare activity) or a resupply and/or waste removal vehicle to supplement the Shuttle.

Information Processing

As the number of space missions grows and their length and complexity increase, the need for Operational Medicine to maintain a flexible, computerized data-base management system also increases. Studies conducted under a JSC contract to define the epidemiologically expected disease/event morbidity rates affecting Space Station crew members have shown that the most valid sources of data are the astronaut inflight and ground-based medical records (4). For Operational Medicine to make informed decisions, complete and up-to-date medical information should be readily available in an appropriately encoded fashion to maintain confidentiality. The effects of repeated and prolonged exposures to microgravity and the particular radiation environment for an individual crew member, for example, will need to be evaluated with an adaptable and conveniently accessible information management system. The data should be available to the NASA life sciences community through the information system and, since this data base is a national resource, to all life sciences investigators through appropriate arrangements. The information management system should be flexible enough to allow for real-time data entry during missions. This would make trend analysis of physiologic parameters possible during prolonged missions on a group or individual basis. Useful information may become available in this fashion that could assist in the development of individualized countermeasures to combat the negative physiological effects of microgravity.

There is room for significant progress in development of a computer-assisted medical decision-making system to supplement the HMF. A microprocessor-based "free text decision support system" as developed under KSC direction is a notable start (5). Also, the work NASA has supported at the University of Maryland to investigate a "computer-based noninvasive physiologic evaluation system" represents significant progress in this area (6). An interactive, intelligent system should eventually be tied into the life sciences medical data-base management system to be updated in real time, so that the most up-to-date information is available for decision making.

Space Medicine Specialist Training

The inclusion of a physician on board to maintain and monitor crew health, diagnose and treat medical problems, and collect medical data will be justified as missions lengthen and crew sizes enlarge. Operational Medicine should define the baseline requirements and determine the educational credentials and training necessary for this medical specialist. While receiving NASA training, the physician could maintain clinical expertise by attending medical conferences and, more importantly, by proceeding through a designated number and type of hospital- and/or clinic-based training/refresher programs involving direct patient care. The particular distribution of specialty rotations should be determined by Operational Medicine based upon the individual's background, interests, and NASA's planned missions. At present, however, a program does not exist for training crew physicians.

Programmatic Issues in Support of Advanced Manned Missions

Operational Medicine must identify the programmatic changes necessary to support long-duration missions staffed by relatively large and heterogeneous crews. To supplement NASA Operational Medicine flight surgeons, a board of specialists will need to be assembled as on-call consultants who represent varying areas of expertise and have been additionally trained by NASA in problems unique to space medicine (7). Operational Medicine is responsible for ensuring that all ground-based medical personnel in support of missions are adequately trained.

In the future, the responsibilities of Operational Medicine personnel will be to support simultaneous missions, which may include missions in low-Earth orbit, geostationary orbit, lunar, or Mars flights, all with heterogeneous crews.

Applied Research of Operational Significance

Operational Medicine conducts research primarily through the Detailed Supplementary Objective (DSO) program of the JSC Space Biomedical Research Institute. This program accepts research proposals having direct relevance to significant operational problems from either intramural or extramural sources, but it does not solicit proposals. All DSO's undergo peer review conducted by the Universities Space Research Association (USRA). Biomedical problems with operational significance studied to date include space motion sickness, cardiovascular deconditioning, pharmacodynamics, and anti-orthostatic countermeasures.

Operational Medicine is involved in a longitudinal study of all astronauts. As part of this effort, the program is conducting yearly physical examinations of present and past astronauts at JSC in an attempt to identify the long-term medical effects of repeated exposures to the space environment (8).

Specific Medical Concerns

Advances in the practice of medicine are dependent upon progress in biomedical research; this principle also applies to space operational medicine. Therefore, much of the material reviewed here concerning prevention, diagnosis, and treatment has been addressed in other reports, such as *A Strategy for Space Biology and Medical Science for the 1980s and 1990s* (NAS, 1987), as well as in the summary of biomedical research given in this report (9). Operational Medicine must maintain close ties with the biomedical research community in order to foster and encourage investigations in critical operational areas.

Prevention. Preventive medical measures are utilized before as well as during a flight. Current crew selection and subsequent retention criteria have proven effective in preventing a number of potential medical problems. This is evidenced by the fact that no U.S. space mission has been curtailed or canceled as a result of inflight medical problems. Consideration should be given now to any special modifications of the crew selection and retention criteria needed to ensure the success of longer missions in the future, such as a Mars mission. The psychological implications of extended missions with longer isolation times will become increasingly important, as will a better understanding of factors influencing group dynamics (10,11).

Before the start of long-duration missions, attempts should be made to identify organic or psychological health problems that could threaten the mission. For example, whole-body magnetic resonance imaging should be considered to screen for occult tumors. Also, based upon a probabilistic model of radiation exposure, a crew member may be encouraged to store bone marrow for autologous bone marrow transplant, should that become necessary. To maximize the fidelity of such a model, every effort must be made to measure the relevant radiation environment accurately.

Operational Medicine has a prime responsibility for inflight occupational health issues, including monitoring the environment and the crew's response to the environment. Detection, identification, modification, and adherence to the limits of spacecraft maximum allowable concentrations (SMAC) for toxic atmospheric contaminants become more important with longer missions, as indicated in "Systems Engineering." SMAC standards need periodic review and revision based on new experience and data. The microbiological atmospheric and surface environments should be continually monitored in a longitudinal manner during long-duration space flight. In addition, the pathogenicity of spacecraft flora in relation to any possible alterations in the host immune system should be evaluated (12). Furthermore, environmental factors, such as temperature, humidity, odor, noise, electromagnetism, and vibration, require scrutiny with respect to health and performance.

Another high priority item for preventing potential inflight medical problems is perfecting the development of the high-pressure extravehicular activity (EVA) spacesuit. The current suits are pressurized at 4.3 pounds per square inch (psi). The ambient Shuttle pressure is 14.7 psi. To avoid decompression symptoms or bends, prolonged periods of prebreathing 100-percent oxygen are mandatory. This procedure, however, has a major impact on flight operations and still leaves a risk of decompression sickness. The development of a high-pressure EVA suit will, therefore, obviate the need for extended prebreathing and significantly reduce the risk of decompression sickness (13).

As spacecraft and missions become more complex, human factors issues, including the design of efficient, compatible human-machine interfaces, become critical to crew safety, satisfaction, and performance. Standards for allowable recreational and personal time consonant with mission requirements will need careful attention. For long-duration missions, development of interpersonal relationships among the crew needs particular consideration. Crew members who will be participating in such missions should be trained in communication skills and in techniques for resolving interpersonal conflicts.

Dietary requirements for prolonged space flight must be established. Any possible dietary manipulations that may help prevent deleterious physiological alterations induced by space flight should be fully explored. To aid in this process, attention should be focused upon developing innovative methods (such as identification of radio-labeled or naturally occurring markers) for dietary monitoring to determine nutritional and/or physiologic status.

Development of countermeasures to the known deleterious physiological effects of space flight requires continued effort from Operational Medicine. Exercise has been shown to prevent some of the cardiovascular and muscle deconditioning known to occur in space-flight. However, it remains to be determined whether the negative calcium balance associated with exposure to microgravity can be effectively reversed by some form of exercise; this issue deserves a high priority effort. Success criteria for a given countermeasure to a space-flight adaptation should be carefully specified. Issues requiring careful analysis are the extent to which physiological adaptations induced by microgravity require countermeasures and the timing of such interventions during the mission.

In general, it is desirable to develop countermeasures that are as simple as possible. Consideration of pharmacological interventions, a human-rated variable-gravity facility, electromagnetic musculoskeletal system stimulation, active electromagnetic radiation shielding, and other possibilities are all less desirable solutions. They should, however, be pursued at least theoretically until the other, more conservative countermeasures have been fully evaluated.

Diagnosis. The development of design considerations for the Space Station HMF, under way at JSC, represents a significant diagnostic effort. An additional challenge will be the diagnosis of disease during space flight for individuals who have developed altered physiological parameters as a consequence of exposure to the unique environment of space.

Treatment. Considerable work has been accomplished in planning for treatment of crew members using the HMF. Future decisions rest upon the results of epidemiological studies of inflight experience and biomedical research. Particularly important areas include the study of incidence figures for specific medical problems that may occur and the evaluation of any potential effects of space flight upon, for example, pharmacodynamics, drug interactions, and wound healing, as in soft tissue versus bone. Decisions concerning the need for specialized hardware, such as a hyperbaric treatment facility, a miniature lithotriptor, or a human-rated variable-gravity facility, must be made by Operational Medicine using accumulated inflight biomedical data. The treatment capabilities that Operational Medicine defines as requirements are anticipated to evolve as the application changes from the Space Station with a CERV, for example, to a manned Mars mission, which will require more autonomy.

Recommendations

Previous sections of this summary detailed areas in which Operational Medicine must make continued progress to ensure the success of long-duration missions with humans in space. Strategies for several high priority areas are outlined below.

Inflight Health Maintenance Facility

- **The Space Station HMF should be designed for flexibility and the capability to change as new experience dictates. The effects of a Space Station CERV** upon HMF requirements should be delineated quickly in case design modifications are necessary for the HMF.

- A long-term goal for the HMF should be to achieve relative autonomy of operation.
- Operational Medicine should define levels of health risk acceptable to NASA.

Future Manned Spacecraft

- NASA should implement development of a CERV for the Space Station as soon as possible.
- The NASP office should assess the biomedical aspects of the NASP design and establish channels of communication with Operational Medicine.

Information Processing

- Operational Medicine should develop and maintain an automated medical information management system before the Space Station is occupied. Pertinent medical data from all astronauts should be available through this system, including both inflight and ground-based longitudinal data. Once long-duration missions are in progress, this system should be updated with inflight medical data collected in real time.
- Operational Medicine should develop a computer-assisted decision-making system as a supplement to any HMF. Ideally, this system should be capable of using the continually updated medical data base.

Space Medicine Specialist Training

- A physician should be included on all long-duration missions.
- Operational Medicine must establish a training program tailored to inflight medical specialists, as well as all ground-based physicians involved in the projected effort. The training should include the use of HMF equipment and embrace sufficient indepth clinical experience to ensure competency in both health care monitoring and emergency medical situations. A curriculum in space medicine with required continuing medical education credits should be established for physician astronauts and flight surgeons at an appropriate medical center.

Programmatic Issues in Support of Advanced Manned Missions

- Operational Medicine should define in detail the personnel requirements necessary to provide medical support for long-duration and/or simultaneous missions (such as Shuttle, Space Station, and Mars missions). In addition, the program should address the use of on-call medical specialty consultants.

Applied Research of Operational Significance

- Relevant biomedical data from astronauts should be obtained at every opportunity, both during flight (at regular intervals during long-duration missions) and longitudinally on the ground. This information should be included automatically in the medical information system in real or near real time. Fail-safe mechanisms should be in place to ensure that the data are complete, accurate, and reliable. It is critical that these data be collected simultaneously on an appropriately matched control population.

- Operational Medicine at NASA should work with other Agency divisions and Government health agencies that deal with issues of mutual interest, such as osteoporosis, radiation exposure, and exercise physiology. Avenues of communication for the exchange of ideas and research results should be encouraged within NASA and among NASA and other organizations and investigators.

Specific Medical Concerns

- A prospective, long-term study should be pursued investigating screening techniques, such as whole-body magnetic resonance imaging or possibly positron emission tomography, for use in crew selection for multiyear missions.

- Development of a high-pressure EVA hard suit for the Space Station should be actively pursued.

- Research into the development of countermeasures for space adaptation, including exercise, diet, and variable gravity, should continue to be pursued with vigor.

- Operational Medicine should periodically review and evaluate environmental standards for spacecraft as an iterative process to ensure crew health and safety.

- Standards for crew recreational and leisure time should be established to maximize crew productivity during extended missions.

Reference List

1. Space Station Projects Office. July 14, 1986. *Medical Requirements of an Inflight Medical System for Space Station.* JSC 31013. Houston: Johnson Space Center.

2. Medical Sciences Space Station Working Group. February 1984. *Space Station Medical Sciences Concepts.* NASA TM 58355. Ed. John A. Mason and Phillip C. Johnson, Jr. Houston: Johnson Space Center.

3. Sulzman, Frank M. January 8, 1987. *Advanced Missions with Humans in Space.* Presentation made to Dale M. Myers, NASA Headquarters, Washington, DC.

4. Lathrop, George D. 1987. *Status Report: An In-depth Epidemiological Analysis of Expected Disease/Event Morbidity Rates Affecting Space Station Crew Members.* NAS 9-17200. Submitted on March 21 to Krug International, Life Sciences Division, Houston.

5. Grams, Ralph R., and James K. Massey, Principal Investigators. February 28, 1987. *The Clinical Practice Library of Medicine (CPLM): An On-line Biomedical Computer Library Final Report.* NASA Grant NAG10-0028. Gainesville, FL: Medical Systems Division of University of Florida, Gainesville.

6. Siegel, John H., Principal Investigator. Date not given. *Medical and Surgical Evaluation of Care of Illness and Injury in Space.* NAS 9-17186. College Park, MD: University of Maryland.

7. Ostler, D.V., and A.M. Shinkman. July 25, 1986. *Systems Requirements Document for HMF Ground Network Nodes.* Version 2. Houston: Johnson Space Center.

8. Mosely, Edward C., Principal Investigator. October 1, 1987. *Longitudinal Study of Astronauts.* NASA-JSC 199-11-21-17. Houston: Johnson Space Center.

9. National Academy of Sciences. Space Science Board. Committee on Space Biology and Medicine. 1987. *A Strategy for Space Biology and Medical Science for the 1980s and 1990s.* Washington, DC: National Academy of Sciences.

10. NASA/National Science Foundation. 1987. *Abstracts from Symposium, The Human Experience in Antarctica: Applications to Life in Space, August 17-19,* in Sunnyvale, CA.

11. Boeing Aerospace Co. National Behavior Systems. October 13, 1983. *Space Station Nuclear Submarine Analogs.* Granada Hills, CA: Boeing Aerospace Co.

12. Biomedical Laboratories Branch. Medical Sciences Division. October 1986. *Space Station: Infectious Disease Risks.* NASA-JSC 32104. Houston: Johnson Space Center.

13. Waligora, James M., David Horrigan, Jr., Johnny Conkin, and Arthur T. Hadley, III. June 1984. *Verification of an Altitude Decompression Sickness Prevention Protocol for Shuttle Operations Utilizing a 10.2 Psi Pressure Stage.* NASA TM 5829. Houston: Johnson Space Center.

J. William Schopf, Ph.D.
Chairperson

Arthur W. Galston, Ph.D.

Keith L. Cowing
Staff Associate

Gravitational Biology

Gravity, an obvious and major environmental factor on this planet, has played a signal role throughout the history of life on Earth. Space-based research provides the opportunity to alter the influence of gravity by exposing organisms to fractional gravity levels ranging from essentially zero up to 1 gravity (g). This exposure allows investigation of the effects of gravity on numerous aspects of living systems. Although weightlessness can be artificially created for tens of seconds in parabolic airplane flights, a prolonged state of weightlessness (or, more accurately, "microgravity") can be achieved only during space flight.

The Gravitational Biology Program is managed at NASA Headquarters. Intramural research is conducted at Ames Research Center, Johnson Space Center, and Kennedy Space Center. As with other areas of life sciences research, extramural research is carried out at a number of universities and research institutions.

The goals of this program, as stated in the *1986-87 NASA Space/Gravitational Biology Accomplishments* (NASA TM 89951), are as follows: "to use the unique characteristics of the space environment, particularly microgravity, as a tool to advance knowledge in the biological sciences; to understand the role of gravity in the biological processes of both plants and animals; and, to understand how plants and animals are affected by and adapt to the space flight environment, thereby enhancing our capability to use and explore space."

A number of recent assessments of space science and technology have noted the importance of space life sciences research in general and gravitational biology in particular. The 1987 report of the National Academy of Sciences' (NAS) Committee on Space Biology and Medicine, *A Strategy for Space Biology and Medical Science for the 1980s and 1990s,* suggested four major scientific goals that should be addressed by a balanced space life sciences research program. Among these is "to understand the role that gravity plays in the biological processes of both plants and animals." The overall program suggested by the NAS committee addresses both basic and applied research combined with an integrated program of ground- and space-based investigations. The committee recognized that inflight centrifuges would be "essential instruments for the future of space biology and medicine" and recommended that "a variable force centrifuge of the largest possible dimensions

This mockup of a 1.8-meter centrifuge is undergoing tests at NASA's Marshall Space Flight Center. The centrifuge was designed at Ames Research Center for use on the Space Station.

be designed, built and included in the initial operating configuration of a [dedicated] life science laboratory."

The National Commission on Space, in its 1986 report *Pioneering the Space Frontier*, recommended seven goals to be pursued in order to "foster [an] integrated approach to research on fundamental questions in science." One recommended goal is to conduct "new research into the effects of different gravity levels on humans and other biological systems." The commission emphasized the need for such research both to "resolve fundamental questions" and to "solve pacing [operational] problems that depend on gravity." A particular need was seen for "long-duration studies of the reactions of humans and plants to the microgravity of free space, the one-sixth gravity of the Moon, and the one-third gravity of Mars." To help accomplish these goals, the commission recommended the "early availability of a dedicated variable-G research facility in Earth orbit to establish design parameters for future long-duration space mission facilities."

Astronaut Sally Ride's report to the NASA Administrator, *Leadership and America's Future in Space*, recognized that "life sciences research is...critical to any program involving relatively long periods of human habitation in space." This report states that "research must be done to understand the physiological effects of the microgravity environment" and "to develop measures to counteract any adverse effects." These efforts were considered crucial to conducting two of its four major goals: a manned outpost on the Moon and a human mission to Mars.

The NASA Advisory Council's Space and Earth Science Advisory Committee, in its 1986 report *The Crisis in Space and Earth Science*, noted that life sciences research activities would "form a major part of the space station" and "because of the unique ties of [this] discipline to the manned program, particular care will have to be taken in the design of the experimental requirements if the promise of [this] field is to be realized."

Scientific Issues

Much of the research conducted by the Gravitational Biology Program is directed toward problems in basic science. Since a number of questions crucial to a detailed understanding of the effects of gravity can be addressed only aboard orbital spacecraft, it follows that NASA, being chartered to contribute to "the expansion of human knowledge of phenomena in the atmosphere and space," should be involved in these endeavors (NASA Act of 1958, Section 102[c][1]).

The program is currently investigating four major topics: Cell Biology, Gravitational Perception and Sensing, Developmental Biology, and Biological Adaptations. A 1-g centrifuge has rarely been flown to provide a space-based control. Consequently, few results noted thus far in space flight experiments can be unambiguously attributed solely to microgravity rather than to radiation or other possible causes. Much remains to be learned.

Gravitational Aspects of Cell Biology

All levels of biological organization appear to be sensitive to the influence of gravity at forces greater than 1 g. Research in cell biology is designed to measure this sensitivity and to understand the mechanisms by which cells respond to gravity both as individuals and as components of multicellular organisms.

Current research is directed toward a variety of objectives: investigating the effects of gravity upon cell structure, division, differentiation, and metabolism; specifying the causes of observed gravitational sensitivity, whether direct (intracellular in origin), indirect (external or system causes), or a combination thereof; determining whether interactions occur between gravity and other environmental factors, such as light and ionizing radiation; examining whether observed changes in cell structure and function are transient or permanent and whether adaptation occurs; determining the scope and course of readaptation (if any) to 1 g; and investigating the use of inflight analytical techniques, such as cell culturing and flow cytometry.

Specific research under way includes studies of the role of amyloplasts as putative statoliths in plant cells and their role in gravitropism, the interaction of otolithic crystals and individual cells in mammalian vestibular systems that seem to function as bioaccelerometers, the effects of gravity upon newly fertilized eggs and their subsequent embryogenesis, the effects of gravity on hormone production at the cellular level, bone cell turnover and growth, microbial growth and sensitivity to antibiotics, and mechanochemical transduction of information between cells.

Research conducted during space flight has shown that mitosis and cytokinesis in plant cells seem to be affected by space flight, as is evidenced by slowed or inhibited cell division. Whether this is due to the lack of a gravitational vector or is a response to radiation or other possible environmental factors is, at present, uncertain.

Gravitational Perception and Sensing

Plant and animal species have evolved a variety of gravisensory capabilities that allow them to use the Earth's gravitational field for orientation during growth and movement. Exposing research specimens to a range of gravitational environments provides an opportunity to examine how different organisms perceive, sense, transduce, and transmit information and respond to a gravitational field. Researchers can also study the evolution of various gravity-sensing systems, the operation of these systems in microgravity, and the neurological component of gravity sensing.

Plant Gravisensing. Plant responses to changes of the gravitational vector are exhibited by alterations in the location and rate of growth. Flight and ground experiments have shown that electrical and ionic currents are detectable as early responses to gravity and that calcium ions are probably involved in the transduction of a gravitational stimulus. Results from space experiments suggest that plant shoot growth may be directed by both gravity and light, whereas root growth may respond solely to a gravitational force.

Current research efforts are directed toward understanding what occurs at the cellular level in the perception of gravitational fields (with emphasis on the role of calcium and hormonal messengers and of intracellular organelles as gravity sensors), the gravitropic responses in stems and roots (and the role of gravity in apical dominance), and the use of clinostats as a ground-based means of simulating variable levels of hypogravity.

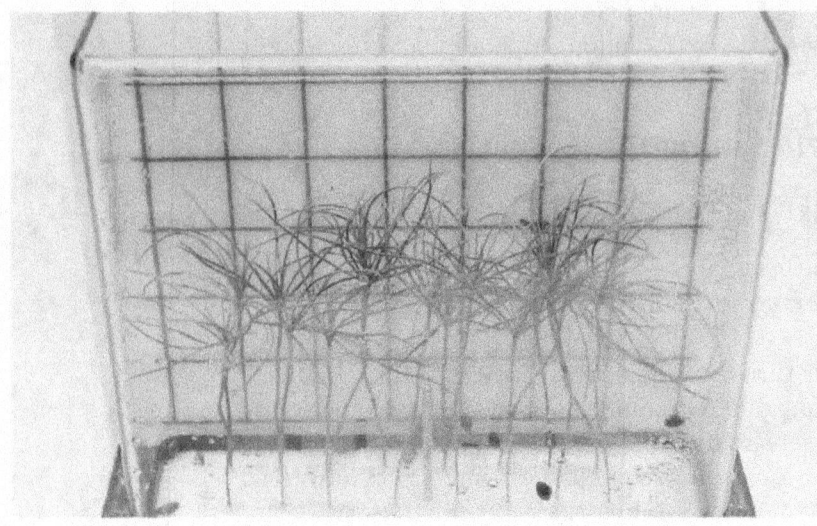

These pine seedlings were flown on Spacelab 2 (STS 51-F), July 29-August 6, 1985, and photographed after the mission. The miniature greenhouses, called Plant Growth Units, allow investigators to monitor the effects of microgravity on the direction of plant growth and on the formation of lignin, a woody substance in the plants that allows them to grow upward against the pull of gravity.

Animal Gravisensing. Animals are capable of sophisticated responses to environmental stimuli by virtue of a complex nervous system integrated with a musculoskeletal system. The Space Biology Program has concentrated on understanding the role gravity has played in shaping the functional organization of animal gravity sensing and organs (bioaccelerometers). Ground-based research focuses on this problem by studying the morphology and physiology of gravity sensors of representative species of animals, both invertebrate and vertebrate, to better understand how gravity sensors process information. Ground-based studies are under way to determine the mechanisms of transduction, including ionic as well as mechanical processes, and of transmission of information from the receptors to the central nervous system. Although work on neurotransmitters and on neural coding is not presently supported, these areas are within the scope of information processing and should be undertaken.

Ground-based research also employs computer-based, three-dimensional reconstruction of gravity sensing and organs of mammals. This research, when combined with results from physiological and neurochemical investigations, can lead to modeling of gravity-sensing systems. Models can be used to predict the adaptive behavior of the sense organs under changed gravitational conditions, such as would be experienced on Mars or in interstellar space. Such theoretical work is useful in singling out the most important questions to ask in space and, therefore, the kinds of experiments that are most critical to conduct on the Shuttle or the Space Station. This kind of computer-based research should be extended to include a comparative series from invertebrates to vertebrates. One reason for doing so is to study the question of whether evolutionary advances from aquatic to terrestrial forms, and from prostrate to upright posture with increased mobility, are reflected in the functional organization of gravity sensors.

The ground-based research under consideration in Space Biology leads naturally to studies in weightlessness. Flight and ground-based experiments have shown that jellyfish rotated on a clinostat (to produce an ambiguous gravity vector) contain reduced numbers of statoliths, suggesting a role for gravity in their normal development. Because animal gravity sensing and organs are functionally organized as weighted neural networks and process information in parallel, they are highly adaptive systems. Some aquatic species possibly adapt quite readily to the space environment because of the buoyancy they experience in their everyday lives on Earth. Terrestrial forms possibly will experience longer periods of adaptation. An unanswered question is whether some species will begin to select for some altered functional organization after multiple generations of exposure to weightlessness, and another is whether progeny of these lines will readily readapt to Earth's gravitational field when returned from their "normal" habitat on the Space Station.

The Effects of Gravity on Organismal Development

As with mature organisms, developing individuals are exposed to a range of environmental factors that exert a strong influence on bodily structure and behavior. The major objective of research in this area is to understand the role of gravity in reproduction, growth, development, and aging.

Developmental Biology of Plants. A few space missions conducted by the Americans and Soviets have carried plant experiments that demonstrated a variety of responses by plants to space flight. The exposure of plants to the space environment seems to alter the character and rate of cell differentiation, accelerating it in some species and apparently slowing it in others. Carrot cells cultured aseptically on defined media develop somatic embryos during space flight as well as on the ground.

The Soviets have grown *Arabidopsis*, a small plant, in space from seeds and brought it through a complete life cycle to produce fertile seeds. As they matured, these plants grew slower, were smaller in size at maturity, and produced fewer leaves and seeds than did ground-based controls. Current research projects include investigation into the effects of gravity on plant cells and embryos; the role of calcium in the regulation of plant development; the genetic basis of gravitropism; the effects of gravity on chromosomes, cell and tissue competence, organogenesis, and developmental timing; and the role of gravity in flowering and fertilization.

Developmental Biology of Animals. The goals of the research program in this area are to determine the effect of gravity on pattern specification in embryonic development; to investigate the involvement of gravity in cell differentiation, histogenesis, organogenesis, and overall system integration during the normal development of organ systems in various vertebrate and invertebrate species; and to learn at all levels of organization the extent to which gravity influences reproduction and the normal development, growth, maturation, and aging of organisms. Current research focuses on such topics as hypergravity and mammalian development, the effects of gravity upon the polarity of amphibian eggs, amphibian development in microgravity, vestibular system development, cytoskeleton formation, the role of gravity in mammalian fertilization and development, and the effect of hypergravity upon the reproductive capabilities of various rodent species. Ground-based studies using hypergravity (centrifuges) and gravity vector randomization (clinostats) are employed to develop techniques and baseline data for studies to be conducted on animals in space. The research will focus on the animals' conception and development.

Biological Adaptations to Gravity

The objectives of this research are to:

- Determine the role of gravity in regulating metabolic rate and products, fluid dynamics, and biorhythms

- Understand the effects of gravity on biological support structures and basic mechanisms of mineral and hormonal metabolism

- Identify the biological effects of the interaction of environmental factors, such as temperature and light, with gravity and determine the mechanisms involved

- Use the space environment as a tool to determine the factors that control the structure and function of organisms.

Animal Adaptations to Gravity. Current research on animal structural adaptations to gravity seeks to determine whether gravity directly affects cellular ultrastructure or exerts its effect extracellularly and to elucidate the mechanism(s) involved. Metabolic studies seek to determine whether temperature regulation is gravity dependent, whether the mechanisms controlling temperature regulation are calibrated for 1 g, and whether normal terrestrial gravity plays a role in establishing basal metabolic rate and biorhythms.

The research program makes extensive use of vertebrate and invertebrate ground-based models to examine the different mechanisms by which life copes with gravity. This is done for three reasons: 1) to study more easily phenomena previously observed only in space, 2) to correlate terrestrial analogs of phenomena seen in space flight, and 3) to provide adequate experimental controls. This research shows that altering the local gravitational field can have a profound impact upon animal physiology. Rats and primates exposed to hypergravity during development experience a modification in their neural thermoregulatory system that causes a delay in their ability to return their body temperature to pre-experimental levels. The biorhythms of growing rats exposed to hypergravity stay depressed for several days and do not synchronize with light/dark cycles as do ground-based control animals. Space flight has also been shown to influence the rat's circadian timekeeping system. Bone-forming cells (osteoblasts) developed by rats flown in space exhibit significantly inhibited rates of bone deposition. Skeletal unloading has been shown to reduce the accumulation of dense, highly mineralized, mature bone, in addition to reducing bone formation. However, if strain is placed upon an unloaded bone via muscle tension, bone growth inhibition is reduced.

Plant Adaptation to Different Gravity Levels. Plant adaptation research is focused on studying the effects of hypo- and hypergravity on metabolism, especially carbohydrate, lignin, and lipid synthesis; on the composition, organization, and size of plant structures; and on fluid dynamics and distribution in plants. Also of interest is to determine how plants respond to the interaction of gravity with various environmental factors, such as light and ionizing radiation, to understand the mechanisms involved, and to separate these effects from those due solely to microgravity exposure.

Findings and Recommendations

On-Orbit Variable-Gravity Facilities

Findings

- On-orbit variable-gravity facilities, which include centrifuges, are required for scientific studies of microgravity:
    - Variable-gravity facilities are needed to isolate the effects of microgravity from all others associated with space flight, including forces encountered during launch and reentry, solar and cosmic radiation, and environmental contamination.
    - Research specimens need to be subjected to different g levels for varying periods of time so that the acute effects of micro- or fractional-g exposure and readaptation to higher g levels can be understood.
    - Long-duration human habitation of the Moon and Mars will require prior long-term studies of the effects of exposure to 1/6 g and 1/3 g on animals and, eventually, humans, including studies of multigenerational exposure to varied g levels.
    - Variable-gravity facilities will play a crucial role as part of a program to evaluate artificial gravity as a potential countermeasure designed to reduce the debilitation caused by prolonged exposure of humans to hypogravity.
- Centrifuge size and design are governed by several factors: the physical limitations imposed by spacecraft volume, the size of research specimens to be

placed within the centrifuge, space required by other research activities aboard the spacecraft, and diameter limits below which problems associated with coriolis forces (the result of an object's angular velocity within a centrifuge) and gravity gradients become intolerable.

Recommendation

- A suite of variable-gravity facilities that include centrifuges of the following designs should be available for gravitational biology research in space:

  — Small centrifuges that can be mounted in middeck lockers, Spacelab/Space Station-style racks, free-flier satellites, or other targets of opportunity for cell biology and small plant research

  — A 1.8-meter diameter centrifuge facility that can be rack-mounted in Spacelab or aboard the Space Station at Initial Operating Configuration (IOC) for rats, squirrel monkeys, and larger plants

  — A tethered (>10-meter diameter) variable-gravity research facility for the Space Station that would greatly reduce coriolis/gradient problems across large animals and that would be operational before the start of any human space missions of extended duration.

Maximizing and Expanding Flight Opportunities

Findings

- Biological research requires the ability to repeat and/or modify procedures based on the results of earlier experiments.

- The lack of flight opportunities for life sciences researchers has numerous unfortunate consequences, including the following: investigators become increasingly discouraged as scheduled flights are repeatedly delayed; graduate students, laboratory space, and institutional support become increasingly difficult to justify; and scientists who had been considering the submission of a proposed experiment decide to pursue other objectives.

Recommendation

- Flight opportunities for life sciences experimenters should be expanded in the following ways so that investigators can repeat or modify trials based on the results of earlier experiments:

  — At least one locker on every Orbiter flight should be reserved and dedicated to life sciences experiments.

  — One Spacelab mission dedicated to life sciences research should be flown as frequently as possible.

  — Autonomous life sciences free-flier satellites should be developed capable of launch on either the Space Transportation System (STS) or an expendable launch vehicle (ELV), STS recovery, or an autonomous soft reentry. Such vehicles should allow a wide choice of orbital inclinations and altitudes, mission length, and scheduling.

— A dedicated life sciences laboratory should be included as part of the Space Station's Permanently Manned Capability (PMC) with emphasis on the use of existing Shuttle, Spacelab, and free-flier hardware.

— U.S./U.S.S.R. cooperation should be continued and increased, especially for Biocosmos missions; and joint U.S./U.S.S.R. *Soyuz/Salyut/Mir* missions should be vigorously pursued.

— Other international collaborative efforts should also be pursued.

— NASA should explore the possibility of using commercial space facilities, such as domestic and foreign expendable launch vehicles, Spacehab, a middeck extension module with middeck lockers that would ride in the payload bay immediately aft of the crew module bulkhead, or the Commercially Developed Space Facility — a Shuttle-launched mini-Space Station with both man-tended and autonomous operating capabilities.

On-Orbit Histology Capability

Findings

- Life sciences research involves the study of dynamic, constantly changing systems. Frequently, the only way to understand adequately the components of these activities is to stop them (by chemical fixation) in serial fashion (both temporally and spatially) and compare the observed changes from one sample to another.

- While the end results of exposure to microgravity are often clear, the interim steps are not. This is especially troublesome when Developmental Biology's concern for the step-by-step progression is considered.

- To date, the ability to perform on-orbit fixation and analysis of specimens has been limited. Most fixation has had to be done post-flight, often after specimens were exposed to the stresses of reentry and landing.

Recommendation

- **The capabilities for manual and/or automated tissue culturing, histology, and light microscopy (all with image-transmitting capabilities) should be included in any life sciences laboratory on the Space Station.**

Computers, Analytical Equipment, and Remote Processing

Findings

- The Life Sciences Division does not take full advantage of recent advances in analytical hardware and procedures.

- To maximize its efforts, the Life Sciences Division needs to incorporate these recent advances into on-orbit and ground-based data analysis.

Recommendation

- The Life Sciences Division should increase the availability and exploitation of its data analysis capabilities in the following ways:
  - On-orbit computer systems should be provided with hardware and software capabilities for telescience, including real-time, two-way (ground/flight) interactive data handling and remote processing.
  - Experimental protocol should be standardized and promulgated whenever possible so as to ease new investigators into the process and allow different/repeated experiments to be more easily compared with research outside NASA.
  - A computerized data base should be established that is accessible through standard online services and that contains space-flight experimental data; a space life sciences bibliography; summaries of past, present, and proposed research; listings of participants and their home institutions; and information on educational opportunities.

Selection of Research Organisms

Findings

- To date, a limited diversity of animals and plants has been flown in orbit.
- Extrapolating results from one species to another is often scientifically inappropriate, leading to incorrect conclusions.

Recommendations

- NASA should use a sufficient number and diversity of taxa to examine representative examples of gravitational perception, sensing, response, and adaptation.
- Adequate plant and animal unicellular and multicellular growth facilities must be provided that include capabilities to rear several generations under automated control. Particular emphasis should be placed on maintaining an in-orbit, multigenerational rat colony.
- Proposed investigations similar in scope to work done previously by other researchers should use standardized experimental plants and animals whenever possible.
- Collaboration between the Gravitational Biology and Controlled Ecological Life Support Systems (CELSS) Programs in the areas of plant research should be encouraged.

CELSS and Gravitational Biology Collaboration

Finding

- Including plants as a component of a CELSS module requires the capability to sustain and maximize plant growth in space. This presupposes understanding of the response of various plant processes to micro- or artificial gravity, a basic goal of Gravitational Biology research.

Recommendation

- **Plant and animal research in Gravitational Biology and CELSS research should be coordinated.**

Reference List

Halstead, Thora W., ed. June 1987. *1986-87 Space/Gravitational Biology Accomplishments.* NASA TM 89951. Washington, DC: National Aeronautics and Space Administration.

National Academy of Sciences. National Research Council. Committee on Space Biology and Medicine. 1987. *A Strategy for Space Biology and Medical Science for the 1980s and 1990s.* Washington, DC: National Academy Press.

National Aeronautics and Space Administration Advisory Council. Space and Earth Science Advisory Committee. November 1986. *The Crisis in Space and Earth Science: A Time for a New Commitment.* No city of publication or publisher given.

National Commission on Space. May 1986. *Pioneering the Space Frontier.* New York: Bantam Books.

Ride, Sally K. August 1987. *Leadership and America's Future in Space.* Washington, DC: National Aeronautics and Space Administration.

U.S. Congress. House of Representatives. Committee on Science and Technology. October 1983. *National Aeronautics and Space Act of 1958, as Amended, and Related Legislation.* 98th Congress. 1st Session. Committee Print.

Arthur W. Galston, Ph.D.
Chairperson

Peter Vitousek, Ph.D.

C. Ross Hinkle, Ph.D.
Staff Associate

Controlled Ecological Life Support Systems

The Controlled Ecological Life Support Systems (CELSS) Program is a NASA effort to create an integrated, self-sustaining system capable of providing food, potable water, and a breathable atmosphere for space crews on long-term missions. Near-term goals are to understand how human life can be maintained in a stable, autonomous system on Earth and in space, and to develop the technological capacity needed to build autonomous life support systems. The system envisioned will depend on a combination of biological and physicochemical processes in which plant primary production is the raw material for human consumption, and vice versa.

A person of average size — about 70 kilograms — requires from 0.5 to 0.6 kilogram of food, 3.0 liters of water, and from 0.75 to 1.0 kilogram of oxygen each day. Laboratory research has shown that these needs could be met by a bioregenerative life support system using higher plants and/or algae. In addition, laboratory estimates indicate that such a system could be effective within the volume constraints of a space vehicle. These laboratory studies need to be verified by a ground-based experimental effort, which would develop design criteria for manned testing leading to a space-based system. This research will take considerable time. **To develop the capabilities that may be required for advanced missions undertaken during the next few decades, the CELSS Program requires significant expansion immediately.**

The sections below discuss the major issues relevant to the CELSS Program. The discussion is based on information elicited in part from presentations given by personnel at NASA Headquarters and the field centers, pertinent scientific literature, and reports by past advisory committees (1,2,3). In addition, the CELSS Study Group collected data by examining published program plans and the published results of CELSS research, by interviewing principal investigators and other researchers, and by visiting Ames Research Center (ARC), Johnson Space Center (JSC), and Kennedy Space Center (KSC).

Issues and Opportunities

The CELSS concept may be viewed as three integrated subsystems: 1) a food production system based on growing plants under controlled conditions, 2) a food-processing system for deriving maximum edible content from all plant parts, and, 3) a waste management system for recovering and recycling all solid, liquid, and gaseous components necessary to support life. Research on the larger system, illustrated in figure 3, is conducted primarily at ARC and KSC (4).

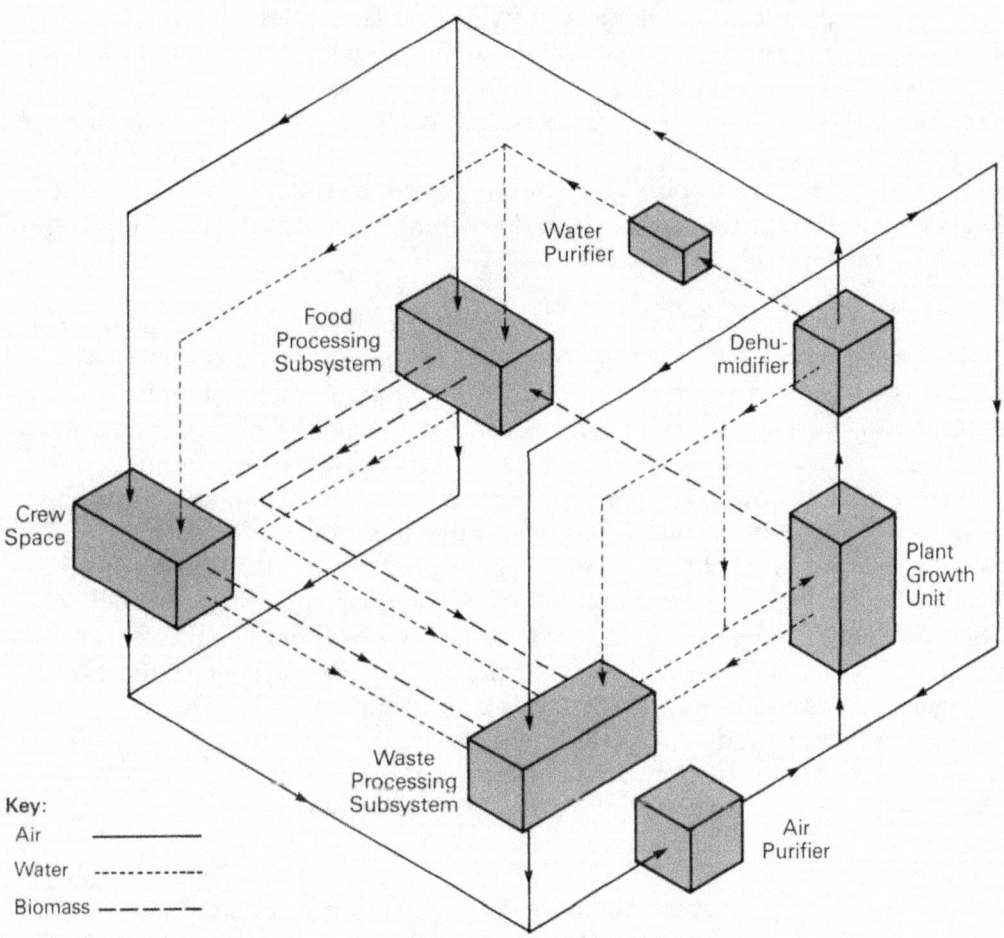

Figure 3. *Schematic CELSS Diagram*

CELSS activities at ARC are directed toward basic research, predominantly on plant productivity. Efforts focus on a number of areas, including increasing the productivity of wheat cultivars, the control of potato tuberization, and the formulation of mathematical models of soybean growth.

CELSS research at KSC concentrates on the Breadboard Project, a pilot effort in CELSS experimentation. The Breadboard facility consists of a >100 m^3 stainless steel tank used to investigate the feasibility of closing the water, oxygen, and carbon cycles, and to evaluate performance of the system. This operation necessarily precedes development of more extensive ground-based manned demonstration units and systems capable of operating in space.

Program Development: The Research Program

The CELSS Program began in 1979 at ARC as a series of workshops that led to a work plan and a series of research grants in 1980. The initial research phase, 1981 to 1985, focused on four major areas. First, research was directed toward increasing food production and processing efficiency and improving growth methods for a range of plants required for a complete human diet. This effort focused partly on the use of algal biomass and a yeast culture to supplement crop plants in the balanced diet and on a continued search for food producing systems more efficient than photosynthesis. Secondly, research on waste processing was directed toward methods to recycle mineral nutrients. Thirdly, work on human requirements was oriented toward optimizing the food production and processing system. Finally, systems research concentrated on developing basic designs and defining the requirements for a pilot study.

Results from the initial studies in increasing food production were very encouraging. By 1984, energy conversion efficiencies of 7-9 percent were obtained with higher plants and 14-18 percent with algae in laboratory studies. The initial studies indicated that a photosynthetically based CELSS could potentially function with as little as 20 m^2 of growing area (20 m^3 of pressurized plant growth volume) and 6 kilowatts of electrical power per person. Subsequent studies continue to support the estimate of 20 m^2, but actual power measurements in CELSS research laboratories indicate that lighting systems and support equipment require from 10 to 15 kilowatts per person. Compared to resupply from Earth, the estimated cost of a bioregenerative system for a 12-person lunar base will reach a break-even point within 5 years at 50-percent food closure and 9 years at 97-percent food closure. Closure of the water and air systems alone results in an immediate advantage when compared to resupply from Earth (5). These high conversion efficiencies coupled with the low area and power requirements, plus the break-even time between resupply and self sufficiency, were considered the basis for starting a pilot study.

CELSS research at ARC is currently focused on maximizing the growth of higher plants under controlled conditions. The plants under study include wheat, barley, soybeans, potatoes, and leaf lettuce. Other research at ARC includes development of a sealed plant-growth facility, using both green *(Chlorella, Scenedesmus)* and blue-green *(Spirulina)* algae as a human food source, controlling algal growth, and evaluating waste processing, especially wet oxidation techniques. Current plans are to continue ongoing research and to request funds for the following: expansion into flight experiments to evaluate crop plant and algal growth in space, and expansion of the evaluation of supercritical wet oxidation processes, cellulose digestion, growth of legumes for CELSS, reclamation, recycling of nitrogen and salts required for plant growth, development of ground-based and flight growth chambers, and development of ecological models of bioregenerative systems. A program plan has not yet been approved for the ARC effort.

Program Development: The Breadboard Project

The KSC Breadboard Project, started in 1985, has three primary goals:

- To establish a NASA capability to integrate and test plant growth subsystems within sealed chambers, including the following:
  - Development of a large, sealed chamber facility and supporting laboratories to grow plants
  - Improvement of methods to grow plants hydroponically under confined space and recycling conditions
- To design, construct, and obtain subsystems to:
  - Control atmospheric contaminants in the sealed chamber
  - Collect and regenerate condensate water and spent nutrient solutions
  - Recycle human and solid plant wastes
  - Process edible biomass
  - Convert inedible biomass to food
- To integrate and evaluate waste management and food-processing subsystems with the biomass production chamber.

A project plan with three phases was completed in March 1986. Phase I (1986-1989) is designed to achieve high performance from plants grown in a controlled chamber. It included the construction of a large (113 m^3, 72 m^3 plant growth volume) biomass production chamber that was tested in an open mode with wheat (December 1986-April 1987) and was sealed during May 1988 for crop production to begin in June 1988. The chamber provides for control of lighting, temperature, moisture, air flow, and collection of all condensate water. Plant growth studies will start with wheat and will advance to multiple crops, such as beans, lettuce, and potatoes, as well as wheat.

Phase II, scheduled for 1987 through 1991, includes the development of subsystems necessary to process food and manage waste. Food processing will involve preparation and conversion of nonedible plant parts to usable materials. Waste management requires the control of CO_2, O_2, and trace gas contaminants; water purification for spent nutrients and condensate; and recycling the constituents of solid and liquid human waste and nonedible biomass. The food processing and waste management systems will be integrated with the biomass production.

The Breadboard facility at NASA's Kennedy Space Center is a large-scale pilot biomass production chamber designed to produce high yields from plants grown in a closed and closely controlled environment.

Phase III (1990-1992) will complete the testing and integration of all subsystems into a true CELSS. Human consumption requirements and waste inputs will be simulated. Atmospheric, biomass, and water cycles will be closed, and all inputs/outputs will be evaluated. Human testing will probably begin about 1995.

Food Production and Processing. Issues that must be resolved in the food production/processing area include:

- Demonstrating large-scale and continuous biomass production using a minimum of space and power
- Finding the optimum balance between plant production and use of biomass to meet human dietary needs
- Testing to determine whether adequate supplies of plants can be continuously produced in microgravity.

Using photosynthesis to convert light to consumable calories has been a main thrust of CELSS research at ARC. Crop plants have been targeted initially as the main mechanism for this conversion. Wheat can be grown at high density (2000 plants m^{-2}), with enriched CO_2 (1000 ppm) and high light (2000 micromol m^{-2} S^{-1}) to produce 56 g m^{-2} d^{-1} at >50 percent edible biomass, thus requiring a plant growth area of 12 m^2 per person. Laboratory studies in potatoes have shown promise, suggesting about 25 m^2 growth area required per person. Demonstration on a large scale with continuous production has not yet been conducted. Verification of laboratory studies is a major thrust of the KSC Breadboard Project, while the ARC research program continues to define optimum conditions for plant growth and photosynthesis efficiency.

Algae and yeast systems have been studied extensively for biomass production, and efficient cultural methods are available for both. Algae are efficient producers, with 14- to 18-percent conversion that is possible during the logarithmic growth phase. Furthermore, 25 percent or more of algal biomass can be extracted as protein, and the cellular content of algae can be controlled by altering growth conditions. For example, stress conditions shift cyanobacterial metabolism favoring increased glycogen production. Thus, algal extracts can contribute to a nutritionally balanced diet. In general, however, it would be desirable to use a combination of higher plants and algae in a CELSS for dietary balance and stability.

At present, we have little experience in growing higher plants or algae under microgravity. Questions concerning the effects of the space environment, particularly of microgravity and radiation, on plant growth and function must be evaluated by flight experiments for CELSS candidate species through several life cycles to determine if a viable stock can be maintained. The NASA Space Biology Program, which conducts microgravity research, has obtained some information on cellular aberrations in microgravity that may be due to radiation effects. It would be useful for this program and the CELSS Program to collaborate on microgravity experiments.

Waste Management. Issues that must be resolved in waste management include:

- Recycling nitrogen from organic wastes and oxidizing the residue to recover CO_2 and H_2O
- Recycling mineral nutrients and H_2O from solid waste and condensate
- Controlling contaminants and atmospheric regeneration.

Although systems exist for recycling nitrogen from organic wastes and oxidizing the residue to recover CO_2 and H_2O, they must be tested within a sealed environment and integrated in the context of the Breadboard Project. Research at ARC to evaluate the efficiency of wet oxidation techniques, including supercritical wet oxidation, will contribute to the choice, testing, and installation of a system for the Breadboard Project. Power consumption and reliability of the system are major concerns that must be resolved.

Recycling of mineral nutrients in the context of a CELSS is not well understood. Ashing of solid wastes, including unused plant biomass and human wastes, will recover the mineral content, but the mineral nutrients must be in a form useful for return to the nutrient solution. Also, sodium chloride (particularly from the human waste) must be separated, perhaps by reverse dialysis, from material going into the plant nutrient solution, to avoid toxic effects on plant growth. Wet oxidation research may help address the problem of nutrient recovery in a CELSS environment.

Currently available technology can recover water as condensate. The procedure will be a major source of water for plant nutrient solutions in the Breadboard Project.

Accumulation of volatile compounds is an important consideration when engineering a sealed environment. Plants generate compounds, such as ethylene, that inhibit plant growth and other compounds, including CO, terpenes, and mustard oils, that are possibly harmful to humans in a sealed environment. Some construction materials liberate gaseous contaminants. The Breadboard Project will be a test bed for monitoring and controlling those contaminants. Initial studies using off-the-shelf technology have indicated that a catalytic conversion system may be adequate for control of volatile contaminants. Maintaining O_2 and CO_2 balance in a sealed system involving plants and man must be demonstrated, but it appears that atmospheric regeneration is feasible in such a system using a combination of biological and physicochemical systems already known. The main requirement is a reliable demonstration in a CELSS.

Human Requirements. Indefinite maintenance of human life requirements within the context of a CELSS is probably feasible, as the Soviets have demonstrated in several closure studies. Issues that must be clearly understood and managed involve provision of a balanced diet, including carbohydrates, fats, proteins for adequate calories, the major minerals and water, as well as essential amino acids, trace minerals, and vitamins. These must be presented as an aesthetically acceptable food product.

Human dietary requirements are understood and could be provided by a vegetarian diet including higher plants and algae. This diet has to be supplied in a CELSS environment with a limited diversity of biomass material and with serious constraints on space and facilities for food preparation. The potential for limited food supplement storage exists, but this must be more fully understood, perhaps at the level of planning meals several months in advance to compensate for plant growth, harvest cycles, and storage life of previously harvested materials. JSC has considerable experience dealing with human requirements, yet its involvement in CELSS-related research has been limited to date.

The aesthetic contribution of plants may be important. Soviet experience has shown that plants are psychologically important for crew morale and positive interactions and that lack of diversity in diet can seriously impair psychological well-being. Critical attention must be given to integration of these considerations into the CELSS research and manned demonstration.

Systems Management and Control. For CELSS to be a success, systems that provide for atmospheric regeneration, food production and processing, and waste management, as well as their control, must be integrated into a reliable system and operated under conditions of reduced gravity. To accomplish these goals, researchers and technicians need to examine monitoring and feedback control systems, automate all systems to reduce human maintenance, establish methods to handle such tasks as maintenance and cleaning, and minimize risk factors and critical failure points. While most of these problems are being investigated at NASA Centers, some contributions have been made by commercial concerns in life sciences, including Boeing, Martin Marietta, and Lockheed.

Through its manned flight program, NASA has demonstrated the capability to handle physical, chemical, and engineering systems. For CELSS to be successful, that same degree of technological capability must be applied to biological systems. Successful demonstration of a pilot-scale CELSS is of paramount importance to accomplishing this complex task.

A Summary of Major Constraints

Limitations on mass, volume, and power are well established in human space flight. Current estimates of 20 m^3 of pressurized plant growth volume and 10 kilowatts of electrical power per person seem reasonable bounds within which a CELSS can operate. Robotics development may solve many of the challenges related to human labor requirements, but serious problems remain in this area.

For crop plants to be used over multiple harvest cycles, a viable seed stock must be maintained. The effects of long-term exposure to the space environment on plant development, growth, and reproduction are not understood. Until adequate long-term flight experiments can be conducted, CELSS will have to be developed with these unknowns.

Little information exists on the reliability of bioregenerative systems or higher plant performance in a closed system. A series of ground-based, integrated CELSS experimental and test systems will be needed to evaluate these issues prior to design of a spaceworthy system.

The Soviet Experience

The Soviets have had a long history of life sciences experiments evaluating the growth of several species of higher plants and microorganisms in space. The experiments include both short-term and long-term (>200 days) studies in space and in combination with ground-based activities involving hermetically sealed Bioregenerative Life Support Systems. The Soviets recognize the need for bioregenerative systems to support long-term space travel, and they have conducted many tests of manned-closed systems using both higher plants and algae in their BIOS programs. A 1-year isolation study with three persons in a hermetically sealed chamber has been completed, along with several shorter term (30 to 50 days) studies in which man-algae and man-higher plant systems were evaluated.

Current European and Japanese Experience

European industrial groups, including Dornier and MBB/ERNO, are conducting CELSS research under sponsorship of the European Space Agency and the German Research and Development Institute for Air and Space Travel. Progress is being made with algal systems and the beginnings of higher plant systems.

CELSS efforts in Japan have been embraced by a community of scientists representing a number of disciplines. Current projects include algal growth in bioreactors, fish-culturing technologies, waste processing, higher plant growth, and related technologies. Although no long-term CELSS program is defined at present, the level of CELSS-related research is expected to increase as the National Space Development Agency of Japan (NASDA) module is prepared for the Space Station era.

Future Directions

How plants perform in space is the "make or break" question for the CELSS Program. Microgravity may cause stress to plants and substantially reduce their productivity, especially when plants must be grown from seed, to seed, to seed in a functioning CELSS. The Breadboard and other CELSS research projects will show what performance can be obtained from plants on the ground, but plant experiments must be conducted in space to test the conclusions of ground-based research.

CELSS flight experiments will require relatively ambitious missions. Even for small, fast-growing plants, the minimum duration of a complete growth cycle is about 45 days. Thus, CELSS experiments will require relatively lengthy stays in space. In addition, CELSS flight experiments will require onboard controls at 1 g to isolate the effects of microgravity from any confounding effects due to radiation, volatile contaminants, or other factors in the space environment. Evidently, CELSS flight experiments will be much more costly undertakings than the Breadboard Project. Therefore, the decision to undertake a flight experiment should wait until the Breadboard Project produces some encouraging preliminary answers, although definition work on space experiments should begin before this milestone is reached. To support a reasonable schedule for the space experiments, the Breadboard Project should be accelerated to provide definitive performance data in about 5 years.

A CELSS plant growth experiment conducted within the Breadboard biomass production chamber.

Both the Breadboard and CELSS flight experiments will need to be followed by further undertakings on the ground and in space as a prelude to a fully functioning CELSS. The Breadboard Project should be succeeded by ground-based tests of a working CELSS with human crews of at least two persons, covering a period of time long enough to evaluate the system's reliability. In space, small plant experiments should be followed by similar experiments using the crop plants identified as optimal in ground-based work. These experiments will require a substantial commitment of pressurized volume for vigorously growing plants the size of soybeans or potatoes. The decision to proceed with this commitment should depend on a relatively firm decision to include a CELSS in an advanced piloted mission. If such a decision is made, the CELSS Program should be ready to begin development of flight-certified hardware for test onboard the Space Station at about the end of the first definition phase of a lunar base or Mars mission.

Findings and Recommendations

Program Requirements

Findings

- NASA may conduct extended human space missions, including a possible manned flight to Mars, early in the 21st century.

  — To make such missions possible, specific criteria for a CELSS need to be established well before the end of this century.

  — The current schedule of CELSS activities, determined largely by budgetary constraints and the time required for new technology development, is inadequate to meet this and other needs.

- ARC, JSC, and KSC have strengths that could be used to support CELSS research under strong Headquarters' leadership.
- The NASA Space Biology Program conducts microgravity research and has obtained information of interest to the CELSS Program, particularly concerning cellular aberrations in microgravity that might be due to radiation effects.

Recommendations

- NASA should plan to develop a fully workable ground-based CELSS within a decade that will provide the basis for designing a flight module and will be integrated into space-based designs.
- NASA should substantially increase funding for the CELSS Program, which is currently budgeted at about $2 million per year. A sizable increase would do the following:
  - Enable parallel rather than sequential development of food production, food processing, and waste management
  - Shorten the time expected to complete the Breadboard Project and to conduct the necessary basic research, systems development, and integration required to provide design criteria for development of a space-operated CELSS.
- NASA should capitalize on the strengths of its field centers in CELSS research.
  - ARC should continue to conduct the basic CELSS research in all areas, with special focus on questions generated by the Breadboard Project.
  - JSC should contribute to subsystem development in the food and waste areas, as well as in manned system testing that would support the Breadboard Project.
  - KSC has initiated and should continue the pilot study in systems integration necessary to establish design criteria for unmanned and manned ground testing, in addition to flight systems.
- The CELSS Program and Space Biology Program should coordinate their activities in microgravity research. Efforts should be directed toward answering questions related to Breadboard Project requirements, such as increasing the efficiency of crop plants, using algal systems, breeding appropriate species for the space environment, and exploring alternative technologies, including tissue culture and genetic engineering.

Flight Experiments

Findings

- A major problem for the CELSS Program is the lack of experience with plants and plant growth systems in space. Many questions can be answered only in the space environment, including:

- The effects of weightlessness on higher plant or algal growth, development, and reproduction
- The capability to grow crop plants from generation to generation
- The capability of plant growth systems developed for a CELSS to function in space
- The effects of cosmic radiation on plant growth and development, including mutations or genetic aberrations.

• Despite the need for experimentation in space, a flight experiment plan for the CELSS Program does not exist.

Recommendations

• **The Life Sciences Division should immediately define and give high priority to the flight experiments needed to resolve key issues pertinent to CELSS.**

• **The Life Sciences Division should continue to pursue every opportunity to fly CELSS experiments on the following:**

— Shuttle and Space Station missions

— Cooperative missions with the Soviets, Europeans, and Japanese

— Vehicles that should be considered for development, such as a dedicated life sciences satellite with the capability to carry extended plant experiments into space and back to Earth.

Integrated and Manned System Testing

Finding

• The Soviets, by constant attention to research over the long run, have gained information about CELSS that the United States does not possess but needs for manned missions of extended duration.

Recommendation

• **Testing with two or more persons in a fully developed CELSS should occur prior to the turn of the century if NASA expects to establish the design criteria to build a spaceworthy module.**

— The tests should be long enough to verify crop/biomass production, waste management, system control and monitoring, and continuous, reliable operation of all systems.

— The CELSS Program should be ready to begin development of flight-certified hardware for testing on the Space Station at about the end of the first definition phase of a lunar base or Mars mission.

Reference List

1. Life Sciences Advisory Committee. November 1978. *Future Directions for the Life Sciences in NASA.* Ed. G. Donald Whedon, John M. Hayes, Harry C. Holloway, John Spizizen, and S.P. Vinograd. Washington, DC: National Aeronautics and Space Administration.

2. Bricker, Neal S. 1979. *Life Beyond the Earth's Environment: The Biology of Living Organisms in Space.* Washington, DC: National Academy of Sciences.

3. Ride, Sally K. August 1987. *Leadership and America's Future in Space.* Washington, DC: National Aeronautics and Space Administration.

4. National Aeronautics and Space Administration. Office of Space Science and Applications. Life Sciences Division. Draft document, November 26, 1985. "Controlled Ecological Life Support Systems (CELSS Program Plan)."

5. Gustave, E., and T. Dinopal. November 1982. *Controlled Ecological Life Support System: Transportation Analysis.* NASA-CR-166420. Moffett Field, CA: NASA Ames Research Center.

Peter M. Vitousek, Ph.D.
Chairperson

Sherwood Chang, Ph.D.

Mathew R. Schwaller, Ph.D.
Staff Associate

Biospherics Research

Human activity in this century has enormously altered the nature of the Earth by changing the landscape and the composition of the oceans and atmosphere. Our perceptions of the changes have become more acute as we have developed the technology to observe human environmental impacts and to document the history of global change. The science of predicting future change, however, remains little developed. The goals of NASA's Biospherics Research Program are to develop and exploit measurement methods and to build quantitative models to predict biological change and the biological consequences of chemical change on regional and global scales.

Issues and Findings

It is not the purpose of this report to define a scientific rationale for biospherics research, a topic covered in detail in numerous other publications, including several identified in the references to this discussion (1,2,3,4). This document focuses primarily on the logistics and policies needed for the Biospherics Research Program to achieve its research goals and objectives.

Scientific Issues

The Biospherics Research Program is the element of NASA's Life Sciences Division devoted to understanding the interaction of biological and global-scale chemical and physical processes. It is a component of a developing international program of studies concerning the Earth on regional and global scales. This program, variously termed the "Mission to Planet Earth," "Global Change," or "International Geosphere-Biosphere Programme" (IGBP), includes scientists from biological, geological, physical, Earth, atmospheric, and marine sciences. It draws much of its impetus from continued observations of human-caused changes in the atmosphere and the realization that these changes may not be reversible for centuries. The NASA Earth System Sciences Committee has described the rationale and some of the major approaches of such an effort in *Earth System Science: A Closer View* (NASA, 1987). It is also discussed in Dr. Sally Ride's report, *Leadership and America's Future in Space* (NASA, 1987).

At present, it is not clear whether NASA will commit its resources to an organized and urgent scientific study of the Earth. Regardless of NASA's decisions, however, it is clear that other domestic and international institutions will increasingly devote their efforts to global studies for the remainder of this century. NASA technology will play a vital role in global research whether or not NASA regards it as a major mission; space science is essential to the study and understanding of global processes. Biologists and Earth scientists can measure such characteristics as forest productivity or ozone concentrations at a particular place and time, but it is only through repetitive, synoptic remote sampling of the land, ocean, and atmosphere from space that these point-measurements can be synthesized into a coherent global picture. NASA led the world in the development of satellite remote-sensing technology for many years; its continued work in this area is essential to the viability of a U.S. contribution to global studies.

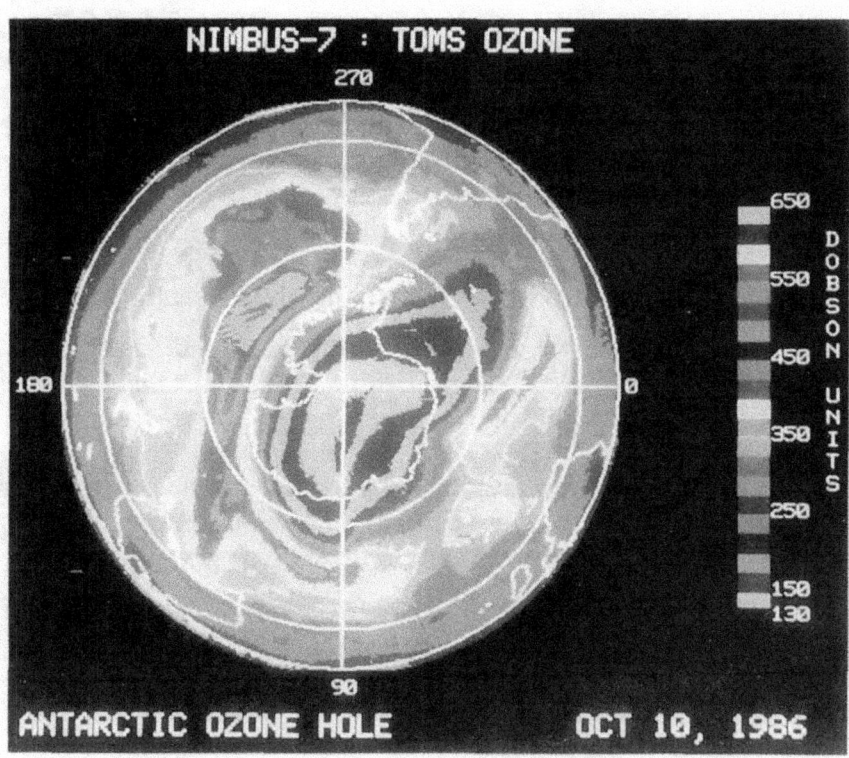

This map of total ozone in the Southern Hemisphere on October 10, 1986, was produced from the Nimbus 7 Total Ozone Mapping Spectrometer. The ozone is the oval feature generally covering Antarctica, portrayed in gray and violet colors. The hole is surrounded by a ring of high total ozone (yellow, green, and brown colors) at middle latitudes.

We believe that it would be counterproductive if NASA's only role in Earth studies were to be a purveyor of images and other data from space-based hardware. In fact, several NASA programs, including Biospherics Research within the Life Sciences Division and Terrestrial Ecosystems and Tropospheric Chemistry within the Earth Science and Applications Division, support interdisciplinary research at the interface between space science and Earth studies. The Biospherics Research Program, in particular, applies NASA technology and modeling skills toward answering global-scale scientific questions. However, the program could do more to develop and exploit that technology. Currently, the aircraft- or space-based technology used by Biospherics Research is developed to specifications that are at times only peripherally related to the requirements of biologists. The program should be involved more substantially in the selection, design, development, and implementation of aircraft- and space-based measurement systems so that these instruments meet the specific requirements of biological research.

The approaches and methodologies developed by the Biospherics Research Program can be applied to a number of issues in addition to understanding biosphere-atmosphere interactions. For example, the Malaria Project proposed to the Life Sciences Advisory Committee by the Biospherics Research Program uses correlation of remotely detectable soil and vegetation characteristics to predict subsequent mosquito populations and to identify areas of incipient malaria outbreaks. The same approach may be applicable to other public health or environmental problems.

Issues of Policy and Infrastructure

Issues of policy and infrastructure address the standing of the Biospherics Research Program within NASA and the institutional capabilities needed to carry out program objectives.

The Problem of Near-Term Data Acquisition. Investigators in the Biospherics Research Program rely on remote sensing as a means for integrating point measurements into a regional or global perspective. At present, however, NASA does not operate any Earth-observing satellites designed for biospherics or related research. Furthermore, there are no plans for any permanently orbiting biospherics-type sensors until the advent of the Earth Observing System (EOS). Although EOS is scheduled for orbit in 1993, the problems of launch vehicle availability and unforeseen budget constraints could delay this mission into the late 1990's or into the next century. Without any active missions, investigators in the Biospherics Research Program are often forced to use data from other missions, from the aircraft program, or from commercial remote-sensing satellites.

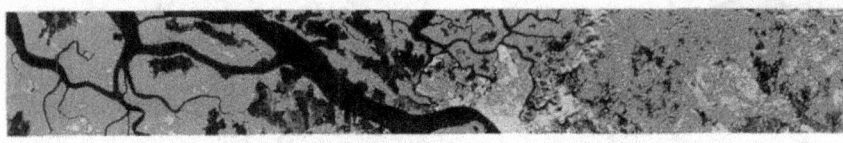

The city of Guayaquil, Ecuador, lies at the center of this picture acquired by the Shuttle Imaging Radar-B on October 7, 1984. Computer processing was used to apply colors to the original image to emphasize various surface features, including the flood plain of the Guayas River, areas of rice cultivation, and forest cover.

Given the lack of suitable orbital missions, remote sensing of the Earth from space has been called "A Program in Crisis" (5). Some investigators in the Biospherics Research Program have maintained measurement and/or modeling efforts by using data acquired from satellites not specifically designed for landscape investigations. Such sources include the National Oceanic and Atmospheric Administration's (NOAA) weather satellites, and promising results have been obtained in application of these data to biospherics investigations. Because of their relatively poor spatial and spectral resolution, however, many experiments simply cannot be conducted with these alternative data sources. Investigators must then rely on aircraft-based sensors or commercial satellites for data. The location, timing, frequency, and scale of data collection from aircraft are limited by high costs and very tight schedules. Commercially acquired satellite remote-sensing data are also costly and often only marginally appropriate for global-scale investigation.

Interagency Cooperation. Even with appropriate data sources, the logistical difficulties of global research pose enormous challenges that transcend the capabilities of any single organization. No single funding agency can sponsor a study of global change to be carried out for decades to centuries, based on topics ranging from the Earth's core to the stratosphere. The research community has concluded that a complex and long-term study of the Earth will require cooperation among many scientific disciplines. The agencies that fund scientific research must develop a similar spirit for national and international cooperation.

The principal domestic agencies that fund Earth science research now include NASA, NOAA, and the National Science Foundation (NSF). NASA has the primary responsibility for Earth science research missions from space, including studies that investigate the Earth as an integrated system. NOAA is responsible for operational weather and ocean satellites and for the development required to improve these capabilities. NSF is responsible for basic research in all areas of Earth and global biological science and plays an especially important role in funding ground-level studies. In addition to these main players, the Department of Energy (DOE), Environmental Protection Agency (EPA), U.S. Department of Agriculture (USDA), and others are developing programs directed to the study of global change. Contacts among these agencies and incipient programs for studying global change have developed rapidly during the period of our Committee review. It is important that NASA remain a committed participant in these efforts and in this area of research.

Funding. The Program Plan for Biospherics Research published in 1983 called for an investigation of global cycles of energy, water, and major biological elements. This strategy was developed by the research communities and has been reaffirmed by the National Academy of Sciences (NAS) (1,2,3), the NASA Advisory Council (NAC) (4), and by special reports to NASA's Administrator (6,7). It is also a substantial part of the current IGBP effort. Unfortunately, NASA's Biospherics Research Program budget for the past 5 years has never exceeded $1.6 million annually, and discretionary funding levels have not been sufficient to support all objectives of the Program Plan for Biospherics Research. Given this situation, the program funds have been distributed among several interdisciplinary research projects. As a central theme, these projects focus on the production of biogenic gases of global importance. They concentrate on modeling and on ground-level investigation of tropical, wetland, and temperate forest ecosystems. The objectives of this approach are to contribute to global biological studies and to maintain a broad constituency of investigators, even if the level of support for each project is relatively small.

The Terrestrial Ecosystems and Biospherics Research Programs. Two programs within NASA have primary responsibility for developing and funding global biogeochemical studies: Terrestrial Ecosystems, which is part of the Earth Science and Applications Division, and Biospherics Research. These programs are responsible for essentially the same disciplines. This has led to some confusion in program management and to the perception by outside investigators that the two programs are competitive rather than cooperative. This perception is not completely justified since the program managers have, on occasion, funded research projects jointly. The perception does, however, have a basis in reality.

A Landsat 4 Thematic Mapper image of the Mississippi Delta reveals striking patterns in vegetation, human cultural activity, and extensive plumes of sediment carried into the Gulf of Mexico.

Insufficient joint planning has contributed to the perception of confusion and competition between Biospherics Research and Terrestrial Ecosystems. A program plan now exists for Biospherics Research; it identifies biogeochemical cycles as a primary research focus. Terrestrial Ecosystems is developing a comparable plan that actively supports research into biogeochemical cycling, as well as other topics. It is logical to expect that overlaps will occur between the most biological portion of the Earth Science and Applications Division and the most Earth-oriented portion of the Life Sciences Division. Without a joint plan to manage such overlaps, the existing confusion and misperceptions will work to the detriment of both programs. Throughout the course of the Life Sciences Strategic Planning Study Committee (LSSPSC) efforts, the two programs discussed a Memorandum of Understanding to define their relative roles and responsibilities. This memorandum, which had not come to resolution at the time this volume was published, represents a positive step for both the Biospherics Research and Terrestrial Ecosystems Programs.

Future Developments

During the course of the LSSPSC effort, the Biospherics Research Program initiated a plan for the Life Sciences Division's contribution to the study of global biology. This plan, known as Project Terra, is intended to use the Earth Observing System for research conducted within the International Geosphere-Biosphere Programme and to conform to the research objectives outlined in *Earth System Science: A Closer View* (4). Project Terra includes research on globally significant ecosystems (such as forests and wetlands in the tropical and temperate latitudes), processes (modeling and measurement of land-atmosphere interactions locally, regionally, and globally), and human problems (the ecological epidemiology of malaria). Project Terra represents a logical step in the evolution of NASA's program in Earth System Science; it should be integrated into the IGBP currently being defined by the National Research Council.

Findings and Recommendations

Findings

- Biologists, ecologists, and Earth scientists can find compelling challenges in *Earth System Science: A Closer View* (NASA, 1988) and *Global Change in the*

Geosphere-Biosphere (NAS, 1986). Representative goals from *Earth System Science* include the following:

— "To obtain a scientific understanding of the entire Earth system on a global scale by describing how its component parts and interactions have evolved, how they function, and how they may be expected to continue to evolve on all time scales."

— "To develop the capability to predict those changes that will occur in the next decade to century, both naturally and in response to human activity."

- International cooperation on global research may be facilitated through the International Geosphere-Biosphere Programme, which was endorsed by the International Council of Scientific Unions. An active commitment to the research strategies of the IGBP would encourage acceptance of the program and would also allow NASA to shape the final form of this program.

Recommendation

- **The Biospherics Research Program should participate in the research goals and challenges outlined in** *Earth System Science: A Closer View* **and** *Global Change in the Geosphere-Biosphere.*

Finding

- NASA does not now operate a permanently orbiting biospherics-type sensor platform, nor are there plans for such a platform until the EOS is deployed in the mid to late 1990's. Investigators in the Biospherics Research Program are thus confronted with high data costs and prospects for severe difficulty in acquiring near-term remote-sensing data.

Recommendation

- **NASA should pursue all viable alternatives, including the following, to bridge the gap in new, remote-sensing data acquisition prior to the advent of EOS:**

    — **Continue to support an aircraft-based remote-sensing program, which is also an important proving ground for experimental sensors.**

    — **Consider the possibilities offered by orbital missions of opportunity, which could include a tropical-areas instrument mounted aboard NASA's low inclination orbiting manned Space Station and remote-sensing devices aboard the Shuttle or on polar-orbiting free-fliers.**

    — **Cooperate with other agencies and organizations where appropriate to design and build satellite sensors and platforms, aid in construction of ground stations at strategic locations for capture of non-NASA data, and/or provide block-grant purchase of data from commercial vendors or from national and international organizations.**

Finding

- Data for Earth System Science will be supplied eventually by the Earth Observing System.
  - EOS consists of four unmanned, space-based platforms housing various remote-sensing devices for Earth observation.
  - It includes two geostationary and two sun-synchronous platforms. Of these, two platforms will be supplied by the United States and one each by the European Space Agency and the National Space Development Agency of Japan.
  - In addition to the orbital platforms, EOS will consist of a ground complement with receiving stations and an advanced data management and data analysis system.

Recommendation

- **The Biospherics Research Program should participate in the design and implementation of the Earth Observing System.**

Finding

- The Terrestrial Ecosystems and Biospherics Research Programs are responsible for similar scientific disciplines within different NASA divisions. Insufficient communication and joint planning has contributed to the perception of confusion and competition between these two programs. Without a plan to manage overlap, the confusion and misperceptions that now exist will work to the detriment of both programs. Cooperation is needed in such areas as determining the research priorities of each program and identifying budget considerations.

Recommendation

- **The Biospherics Research Program should develop a program plan for participation in global biological research that reflects the existence of a large international effort in global research.**
  - **This program can conduct some portion of that global program better than any other group. The program plan should identify and focus on those areas of research.**
  - **The Terrestrial Ecosystems Program should undertake a similar effort; areas of individual and joint interest should be clearly detailed in the Memorandum of Understanding negotiated between these two groups.**

Finding

- The funding requirements and logistical difficulties of global research on a long-term basis pose enormous challenges that transcend the capabilities of any single organization. Cooperation is needed among the agencies that support research in this area, including NOAA, NSF, DOE, USDA, and EPA.

Recommendation

- The Biospherics Research Program should plan a larger role in interagency cooperation, especially among the agencies with interests in global research.

    — Interagency cooperation should be encouraged through NASA's participation in such a body as the Committee on Earth Sciences, under the Office of Science and Technology Policy.

    — Participation should be facilitated by an internal NASA advisory group on Earth System Science, with members drawn from the Program Manager ranks.

    — A clear line of communication should be established between any internal NASA advisory group and the NASA representative to the Committee on Earth Sciences.

Reference List

1. National Academy of Sciences. Space Science Board. Committee on Planetary Biology and Chemical Evolution. 1981. *Origin and Evolution of Life — Implications for the Planets: A Scientific Strategy for the 1980's.* Washington, DC: National Academy Press.

2. National Academy of Sciences. Space Science Board. Committee on Planetary Biology. 1986. *Remote Sensing of the Biosphere.* Washington, DC: National Academy Press.

3. National Academy of Sciences. Commission on Physical Sciences, Mathematics, and Resources. U.S. Committee for an International Geosphere-Biosphere Program. 1986. *Global Change in the Geosphere-Biosphere: Initial Priorities for an IGBP.* Washington, DC: National Academy Press.

4. National Aeronautics and Space Administration Advisory Council. Earth System Sciences Committee. January 1988. *Earth System Science: A Closer View.* Washington, DC: National Aeronautics and Space Administration.

5. National Academy of Sciences. Commission on Engineering and Technical Systems. Space Applications Board. 1985. *Remote Sensing of the Earth from Space: A Program in Crisis.* Washington, DC: National Academy Press.

6. National Commission on Space. May 1986. *Pioneering the Space Frontier.* New York: Bantam Books.

7. Ride, Sally K. August 1987. *Leadership and America's Future in Space.* Washington, DC: National Aeronautics and Space Administration.

Sherwood Chang, Ph.D.
Chairperson

J. William Schopf, Ph.D.

Mitchell K. Hobish, Ph.D.
Staff Associate

Exobiology

Exobiology is an interdisciplinary program of scientific research conducted by the NASA Life Sciences Division, located within the Office of Space Science and Applications (OSSA). As its goal, the program seeks to understand the origin, evolution, and distribution of life in the universe. Just as cosmic evolution implies that all matter in the solar system had a common origin in a cloud of interstellar gas and dust, so does biological evolution imply that all organisms arose by divergence from a common ancestry. Thus, life may be viewed as the product of countless changes in the form of primordial matter wrought by the processes of astrophysical, cosmochemical, geological, and biological evolution that are integral aspects of the development of the universe. From the knowledge gained in this program, it will become possible to formulate a general theory for the natural origin and evolution of life in the universe.

Through both ground- and space-based research, the Exobiology Program seeks answers to these prime questions: How did the development of the solar system lead to the formation and persistence of habitable planetary environments? How did life originate on Earth? What factors operating on Earth or at large in the solar system influenced the course of biological evolution from microbes to intelligence? Where else may life be found in the universe? These questions hold great interest for both scientists and the general public, addressing as they do the history and possible uniqueness of life on Earth and prospects for its existence and detection elsewhere in the galaxy.

To answer these questions, specific research goals and objectives have been identified for the six components that comprise the Exobiology Program. Attainment of these goals and objectives will elucidate the evolutionary pathway followed by the major elements that make up living systems — the biogenic elements — leading from their origins in stars, through the formation of the solar system and planets, to the origin and evolution of life on Earth and elsewhere. Accordingly, the broad scientific scope of the program is organized into four evolutionary epochs corresponding to major stages in the evolution of living systems and their chemical precursors: 1) Cosmic Evolution of the Biogenic Compounds: The growth in complexity of the biogenic elements from nucleosynthesis in stars, to interstellar molecules, to organic compounds in asteroids and comets; 2) Prebiotic Evolution: In the context of planetary environments, the development of the chemistry of life from simple components of atmospheres, oceans, and crustal rocks, to complex precursors, and to cellular life forms; 3) Early Evolution of Life: Biological evolution from the first organisms to multicellular species and the relationship of biological evolution to planetary evolution; 4) Evolution of Advanced Life: The emergence of advanced life forms as influenced by planetary or astrophysical phenomena. Two additional components support these epochal program components: 5) Solar System Exploration and 6) Search for Extraterrestrial Intelligence (SETI). The former carries the search for evidence of life or its chemical precursors to other bodies in the solar system with spaceborne instruments and experiments; the latter scans the skies with ground-based radiotelescopes for signals produced by technological civilizations in the galaxy.

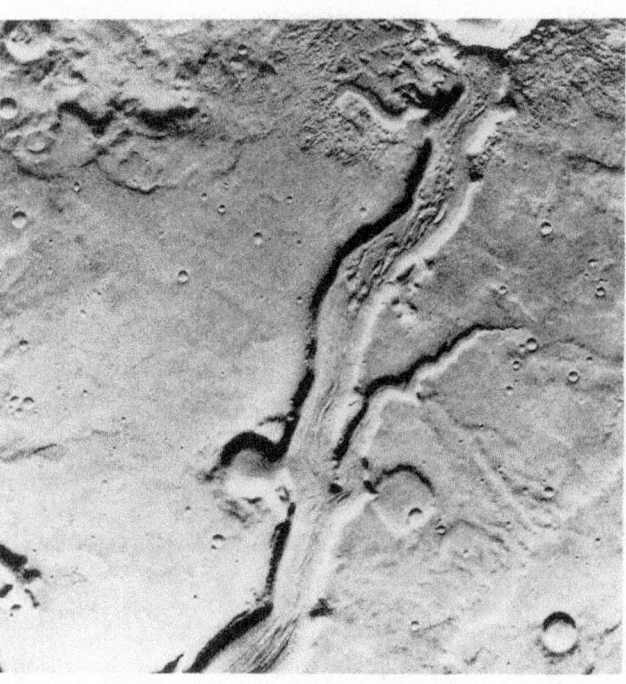

Taken by Viking Orbiter 1, this photograph of the Martian surface shows a small channel system. The channel, about 2.5 kilometers in width, has flow features along its length and tributaries that join the main channel. This and similar channels on Mars suggest that water erosion may have occurred during a warmer and wetter epoch in the planet's history.

Exobiology and NASA's Charter

Among all Federal agencies, NASA is uniquely chartered to explore the matter that exists and the processes that occur in space within the solar system and beyond. The present understanding of biology and the natural history of life on Earth leads to the conclusion that life originates and evolves on planets and that biological evolution is subject to the vicissitudes of planetary and solar system evolution. For these reasons, unparalleled opportunities to contribute to the program's goal are embodied in the missions and projects associated with NASA's exploration of space; the rationale for conducting the Exobiology Program in NASA has been and continues to be firmly rooted in the Agency's charter.

Scientific Goals and Objectives and Strategies for Achievement

Results of research supported by NASA's Exobiology Program show that water and the prebiotic organic compounds believed to have been required as the building blocks of the chemical precursors to living systems are widespread in the solar system and beyond. The ubiquity of these compounds forms the basis for the hypothesis that the origin of life is inevitable throughout the cosmos wherever these ingredients occur and suitable planetary conditions exist. Given the enormity of the observable universe, a prediction originating from this hypothesis is that extraterrestrial life is widespread.

Testing the theory that life is a natural consequence of the physical and chemical processes engendered by the evolution of the cosmos requires a broadly based, scientifically rigorous, and well-coordinated program of research. Implementation of such a program may be expected to yield major advances toward elucidating the following: 1) the relationship between the organic matter of interstellar clouds and primitive solar system bodies, such as comets and asteroids, and the processes of prebiotic evolution on Earth that led to living systems; 2) the bounds placed on the origin of life by the physical and chemical conditions associated with the formation and early evolution of planets; 3) geochemically plausible pathways by which prebiotic chemical systems became living systems; 4) the characteristics of the common ancestor of extant life and the conditions that prevailed in its environment; 5) the presence of extant or extinct life on Mars; 6) the influence of geological processes (such as tectonism) and astrophysical processes (including asteroid impacts) on the evolution of life; and 7) the probability that technological civilizations exist nearby in the galaxy. The scientific goals and objectives and the strategies for achievement described below are consistent with those given in the draft report of the Space Science Board's Committee on Planetary Biology and Chemical Evolution (1).

The Cosmic Evolution of the Biogenic Compounds

The primary research goal in this program component is to determine the pathway followed by the principal biogenic elements (C, H, N, O, P, S) and their compounds, including water, from their birth in stars to their incorporation and final transformations in the asteroidal and cometary building blocks of planetary bodies. Six stages along the pathway of cosmic evolution have been defined for study: 1) nucleosynthesis and ejection of biogenic elements and compounds into the interstellar medium (ISM), 2) chemical evolution in the ISM, 3) protostellar collapse, 4) chemical evolution in the protosolar nebula, 5) growth of planetesimals from dust, and 6) accumulation and thermal processing of planetoids (2).

The possibility of determining a cosmic history for the biogenic elements and compounds is becoming a reality as exciting discoveries emerge from new astronomical observations of the ISM, increased theoretical understanding of processes occurring in the ISM and during formation of the solar system, and detailed analyses of meteorites, comets, and cosmic dust (2,3).

Organic compounds containing up to 11 atoms of H, C, and N have been detected in the gas phase of interstellar clouds along with many simpler compounds that, in the context of the chemistry of early Earth, have been attributed as building blocks in the prebiotic synthesis of amino acids and nucleotides. Organic matter also appears to be a major component of interstellar dust. And around carbon stars are seen hydrocarbons and fine-grained carbonaceous dust presumably formed from elemental species flowing out of the interiors. Along with water in the forms of ice and gas, organic compounds are widespread in the galaxy in interstellar and circumstellar regions, thus supporting the view that the chemistry of the cosmos is largely organic chemistry. Many tantalizing questions are raised by these astronomical observations. For example, what level of molecular complexity can be attained in circumstellar and interstellar organic chemistry? Are amino acids or nucleic acid bases formed in the ISM?

How does water interact with organic and inorganic matter in the interstellar medium? What gas and dust species survive transit from their circumstellar site of origin to the ISM? Progress in obtaining answers to these and other questions will require astronomical observations with airborne and spaceborne telescopes, as well as theoretical studies aimed at elucidating the relationship between physical conditions in these environments and the nature and abundance of observed species.

Astronomical observations and theories of solar system origin indicate that formation of stars and, presumably, associated planetary systems occurs in dense regions of the ISM typically where organic matter and water are seen as molecules in the gas phase and as components of dust grains. The theories suggest, moreover, that physical conditions during evolution of the solar nebula would allow material from the parent interstellar cloud to survive in the outer nebular regions and become incorporated in primitive bodies, such as comets and asteroids.

Studies of Comet Halley have revealed a variety of simple organic compounds and a fascinating, but poorly characterized, complex mixture of higher molecular weight particles, composed only of various combinations of the elements C, H, O, and N. The simple compounds, including hydrogen cyanide and formaldehyde, are among the most abundant in the interstellar clouds, thus underscoring the probability that comets contain components of interstellar origin. Establishing such origin will require automated spacecraft investigations of comets, as planned for the Comet Rendezvous Asteroid Flyby (CRAF) Mission, and detailed analysis in terrestrial laboratories (see below) of samples returned from the nucleus of a comet, as envisioned for the Rosetta Mission.

Recent analyses of carbonaceous meteorites and cosmic dust revealed that some of the organic matter in them, including amino and carboxylic acids in one meteorite, contains anomalously high ratios of deuterium to hydrogen approaching those seen only in molecules observed in the ISM. Additional research on samples of meteorites and dust needs to be conducted to determine how widespread such deuterium anomalies are among the classes of organic compounds and among types of samples. These studies should also be expanded to seek anomalies in other biogenic elements. Whether the organic matter containing these isotopic anomalies originated in the ISM or was formed as secondary products in the solar nebula or in the asteroidal parent bodies of the meteorites from interstellar chemical precursors is a central issue that remains to be elucidated. To help resolve this question, more laboratory and computer experiments should be undertaken to simulate the chemistry of these environments. These investigations should yield isotopic and molecular structural criteria suitable for use in distinguishing between various mechanisms and environments of formation when applied to data obtained by astronomical observations and sample analyses. Opportunities to carry out some of these experiments under the microgravity conditions of the Space Station should be exploited (2,4).

Carbon-containing minerals of exotic origin have also been isolated from meteorites. On the basis of their contents of isotopically anomalous elements, silicon carbide grains appear to have formed as a condensate in the outflows from carbon stars; and diamond grains are thought to have been either similarly derived or produced in the interstellar medium. That these grains have survived the journey through solar system formation to incorporation in the parent bodies of the meteorites establishes another link between astrophysical events that predated the solar system and the accretion of primitive objects that must have occurred early in Earth history. Additional phases linking specific stellar origins to solar system material should be sought in samples of asteroidal or cometary origin.

Although great uncertainty continues about how and where the molecular species were formed, the existence of complex mixtures of extraterrestrial amino acids, hydrocarbons, carboxylic acids, and many other classes of organic compounds in carbonaceous meteorites is well established. These same 4.5-billion-year-old meteorites are made up of clays, carbonates, sulfates, and other hydrous materials that were produced by the actions of liquid water on preexisting assemblages of anhydrous minerals, thus recording the earliest history of weathering reactions in the solar system. Together, the organic matter and minerals of carbonaceous meteorites suggest that on certain asteroidal bodies, environments existed of the type that may have occurred on the primitive Earth during prebiotic evolution. How widespread these environments were among the asteroids and what factors were responsible for their occurrence are questions of great interest that can be addressed best by exploration of these small bodies by spacecraft.

Findings for Cosmic Evolution of Biogenic Compounds

- Data from astronomical observations of organic matter and water in astrophysical environments and from detailed analyses of samples derived from asteroids and comets are critical to forging the links in the chain of cosmic evolution connecting the origins of biogenic compounds in stars and interstellar clouds to their occurrences in the building blocks of planets.

- Simulations of processes occurring in astrophysical environments conducted both on the ground and under microgravity conditions on the Space Station are needed to elucidate the mechanisms of synthesis and destruction and the limitations on the development of complexity in the organic matter of interstellar clouds, protosolar nebula, comets, and asteroids.

Recommendations

- **NASA should implement its plans for the following missions and facilities so as to provide new opportunities for direct study of the organic chemistry of comets and asteroids, for infrared observations of organic matter in the cosmos, and for the conduct of astrophysical experiments in space:**

  — Comet Rendezvous Asteroid Flyby Mission

  — Rosetta/Comet Nucleus Sample Return Mission

  — Space Infra Red Telescope Facility (SIRTF)

- Space Station Cosmic Dust Collection Facility
- Space Station Gas Grain Simulation Facility.

- **The Life Sciences, Solar System Exploration, and Astrophysics Divisions in OSSA should enhance their support of ground-based interdisciplinary research on the biogenic elements and compounds through ongoing astronomical observations, laboratory and computer simulations of organic cosmochemical processes, and investigations into the origins of biogenic compounds and phases in meteorites and cosmic dust.**

Prebiological Evolution

This evolutionary epoch spans the time from the formation of Earth to the origin of life. The research goals for this period are twofold: 1) to understand what conditions were like on the early Earth at the time of the origin of life and how these conditions developed as a result of planetary processes operating over time; 2) to understand how metabolic and genetic systems originated and were incorporated into primitive cellular life forms under conditions that prevailed on the primitive Earth. Opportunities to seek evidence of chemical evolution or the possible origin of life on other planets will be provided by the Titan/Cassini, Mars Observer, and Mars Rover/Sample Return Missions. Use of the Great Observatories and an Astrometric Telescope or a Circumstellar Imaging Telescope will permit the detection of extra-solar planetary systems. Opportunities to learn about conditions on the prebiotic Earth are also likely to be obtained from missions to Mars, the Moon, and other bodies in the solar system.

A gap in the geological record exists between the formation of Earth 4.5 billion years ago and the oldest rocks, with ages of 3.8 billion years. Fossil evidence from 3.5-billion-year-old sediments indicates the existence of diverse marine microbial ecosystems, thus pointing to the origin of life much earlier in the first billion years of Earth history. The lack of a record for this time means that the environmental conditions must be inferred by extrapolation backward in time from the existing record and forward in time from models of planetary formation and early evolution, or through comparative study of extraterrestrial bodies — Mars, Venus, the Moon, the primitive asteroids and comets, and the satellites of the Giant Planets, especially Titan.

Data obtained from the *Viking* missions to Mars are widely interpreted to signify the absence of extant life. The recently detected fluvial features and apparently layered sedimentary deposits, however, have been attributed to the action of liquid water in the first billion years of the planet's history. These observations indicate that Mars was more Earth-like early on. And they hold open the exciting possibility that life also arose on Mars during a more clement climatic period, but then became extinct as the climate changed. For this reason, samples returned to Earth from Mars would be of enormous scientific value for the Exobiology Program. Not only would their analysis permit a more thorough determination of the possible origin of life on Mars, but they would also be invaluable in helping to fill the gap in Earth's geological record and providing a means of reconstructing early environmental conditions.

In the absence of a geological record, the development of physical models for the formation and early evolution of the terrestrial planets — Earth, Mars, and Venus — is essential to placing bounds on the range of physical and chemical conditions that may have existed during their first billion years of history. The value of these models will lie in their ability to elucidate the properties and processes that endow planets with their inventories of biogenic elements and that govern the composition of atmospheres over time, the history of liquid water, and the nature and distribution of environments conducive to the origin of life.

A highly chemically reduced atmosphere dominated by methane and nitrogen was postulated in early models of the primitive Earth and is exemplified by the current atmosphere of Saturn's satellite, Titan. Many laboratory experiments have shown that most of the biochemical building blocks of proteins, nucleic acids, and membranes can be synthesized under so-called prebiotic conditions starting with these atmospheric constituents and water. Although the *Voyager* flyby missions revealed traces of many organic compounds in Titan's atmosphere, the degree of molecular complexity attained in the atmosphere and the physical processes responsible for their syntheses are unclear. The deployment of chemical probes into Titan's atmosphere and surface, as envisioned for the Titan/Cassini Mission, will shed much more light on these uncertainties. Direct comparisons between laboratory experiments and a planetary atmosphere would provide a unique opportunity to test models for abiotic organic synthesis.

Recent models of Earth that take into account early core formation and subsequent outgassing of an atmosphere during late stages of planetary accretion argue for a thick primordial atmosphere dominated by carbon dioxide overlying a warm ocean. In the few laboratory experiments and computer simulations that have been carried out, abiotic synthesis of organic compounds appears to be difficult to achieve in such an atmosphere. Plausible mechanisms for synthesizing the organic building blocks necessary to construct the first cellular metabolic and genetic systems starting from carbon dioxide, water, and nitrogen in the atmosphere and with the components of seawater have not been extensively studied and remain to be demonstrated for these models.

In this context, the discovery of abundant organic matter in Comet Halley, as well as in carbonaceous meteorites, has underscored the possibility that comets and asteroids may have supplied some of the precursor molecules for the synthesis of biochemical compounds during Earth's early history, in addition to supplying much of the planet's endowment of water and biogenic elements. A quantitative assessment of this question will require knowledge of the inventories of organic compounds in comets and asteroids, determination of the size distribution of comets and asteroids that struck the Earth in its first billion-year history, and a better understanding of the physical and chemical effects of impacts involving medium-sized (tens of kilometers in diameter) to large (hundreds of kilometers in diameter) cometary and asteroidal objects. These studies should also reveal the nature and intensity of the fluctuations in Earth's environments caused by such impacts, which may have had a bearing on the ecological niches available over time for the origin and early evolution of life.

The emergence of models of a prebiological environment rich in carbon dioxide and poor in organic compounds raises the possibility that the first organisms may not have been limited to those depending solely on heterotrophic metabolism, the use of preformed organic compounds for energy and for cellular biosynthesis. Permissive evidence of anaerobic photosynthetic microorganisms dates back to the earliest known fossil record, and sulfide oxidation is a capability that appears among prokaryotic organisms with the most ancient phylogenetic lineage. These considerations suggest that the nature of the earliest bioenergetic and biosynthetic pathways remains an open issue, and both autotrophic and heterotrophic organisms could have coexisted in Earth's earliest biosphere.

Although many studies have been directed toward the synthesis of monomers and oligomers of amino acids and nucleic acids under putative prebiological conditions, relatively few investigations have been carried out to understand how the metabolic function itself arose in the prebiological environment. Research on chemical models of metabolic systems should be intensified, and efforts should be made to develop photo- or chemo-autotrophic systems. The roles of peptides, minerals, and membrane-forming organic compounds in these models should be investigated. The photochemical or geochemical oxidation-reduction reactions that provide the energy sources in these models should be consistent with environmental constraints.

Understanding how a self-replicating system with a genetic code arose on Earth is arguably the central problem in the origin of life (3). The theory that nucleic acids were the first replicating systems has gained considerable strength from the revolutionary discovery that ribonucleic acids (RNA's) are capable of splitting and joining oligonucleotides. In principle, primitive RNA's could have been capable of catalyzing rudimentary metabolic reactions as well as replication. Recent advances in RNA technology make it possible to synthesize sequence-specific RNA's for the purpose of assessing this possibility. Studies on the reactions and catalytic properties of RNA's and RNA-like compounds aimed at development models for molecular self-replication should be intensified; they should include assessments of the limitations placed on these systems by environmental conditions consistent with early Earth models.

Clay minerals have also been proposed as the first replicating systems, but this alternative has received little experimental study. Criteria need to be established to distinguish replicating clays from nonreplicating systems, and laboratory investigations of clay syntheses at low temperatures and the role of organic chelating agents in the syntheses should be initiated to test the clay theory.

The complex contemporary apparatus for translation of the genetic information stored in nucleic acids into protein biosynthesis must have had its beginnings in much simpler processes. At some fundamental level, interactions between nucleotides and amino acids leading to the formation of peptide bonds were essential, and these interactions are likely to have originated during prebiological evolution. Theoretical and experimental evidence should continue to be sought for specific interactions that can be related to codonic or anti-codonic relationships and to the ability of nucleotides to catalyze the synthesis of peptide bonds.

Findings for Prebiological Evolution

- Knowledge of the conditions on the early Earth is essential for the development of physical-chemical models for the origin of metabolic and genetic systems. Major uncertainties persist, however, because a geological record is lacking.

- Mars continues to be the prime target in the search for evidence of prebiological evolution and fossil extraterrestrial life in the solar system.

- Much progress has been made on the synthesis of prebiological monomers and oligomers based on methane-rich planetary atmospheres, but little has been done to assess the possible origins of organic compounds in carbon-dioxide-rich surface environments. The development of model chemical systems capable of metabolic function in either type of environment has been largely unexplored.

- As candidates for the first self-replicating systems capable of both metabolic and genetic function, the potentiality of RNA-like molecules has been heightened by the discovery that RNA has catalytic properties, while the alternative of clay minerals has received relatively little experimental emphasis.

Recommendations

- **Research programs in the Life Sciences and Solar System Exploration Divisions of OSSA should direct theoretical studies of planetary phenomena, such as accretionary impacts and the origin and evolution of the atmosphere, oceans, tectonic regimes, and climate, toward determining the range of physical and chemical conditions that may have evolved during the first billion years of Earth history in the near-surface regions of the oceans and the atmosphere and the hydrothermal environments on land and under the sea.**

- **The Life Sciences and Solar System Exploration Divisions of OSSA should pursue vigorous programs of remote observations, ground-based research, and exploration of extraterrestrial bodies — Mars, Venus, the Moon, the primitive asteroids and comets, and the satellites of the Giant Planets, especially Titan — with emphasis on acquisition and study of samples returned from Mars, to fill the gap in Earth's geologic record and to determine the limitations on prebiological evolution elsewhere in the solar system.**

- **The Exobiology Program should continue on a broad front to support research on the prebiological evolution of functional complexity leading** toward living systems with emphasis on the following areas:

- The organic synthesis of cellular building blocks in the context of carbon-dioxide-rich atmospheric and hydrothermal environments
- The organic and inorganic chemical models for metabolic and self-replicating systems compatible with existing constraints on early environmental conditions
- The nature of interactions between monomers or polymers of nucleotides and amino acids that constitute the physical-chemical basis for the genetic code.

Early Evolution of Life

The research goal for the Early Evolution of Life is to understand the relationship between planetary evolution and the evolution of unicellular life from the origin of the universal common ancestor to the emergence of multicelled organisms. Two avenues are available for study of evolutionary history on Earth: 1) the biological record preserved in the metabolic patterns of extant organisms and the sequences of amino acids and nucleotides in their proteins and nucleic acids, respectively; 2) the geological record of fossil life and its environment preserved in ancient sedimentary rocks. The discovery and study of fossil organisms in ancient sedimentary rocks returned from Mars could yield unique insights into the evolution of extraterrestrial life. Searches for astronomical evidence of disequilibria hold promise of revealing the distribution of life forms beyond our solar system.

Although the oldest rocks of Earth are 3.8 billion years old, the existing record of biological evolution begins only at 3.5 billion years, and rocks containing fossils are very sparse until 2.8 billion years ago. To obtain a more complete geological record of life on Earth, the search must continue for additional unmetamorphosed sedimentary rocks older than 3.0 billion years in terrestrial continental deposits. If life arose on Mars over the same period of time as it did on Earth, the planet's relatively low level of geological activity may have permitted more complete preservation of a record of early biological evolution. The return of samples from ancient Martian sedimentary environments would make available a geological record that could permit the beginnings of a comparative paleontology among planets.

The earliest sedimentary rocks on Earth containing stromatolitic and microfossil evidence of microbial ecosystems have been found in Western Australia and South Africa (3). The sediments appear to have been deposited in shallow marine hydrothermal environments on the flanks of volcanic island platforms during relatively quiescent periods between cycles of volcanic eruptions. Except for the effects of oxygen in the atmosphere today, this early setting resembles in many respects habitable environments that exist currently or that may have existed on Mars during an early period of active volcanism. The laminar structures of the stromatolites, the morphology of the microfossils, and the carbon isotopic composition of the associated organic matter are consistent with the presence of both heterotrophs and autotrophs, with filamentous, phototactic, and probably autotrophic organisms composing the major stromatolite-building microorganisms in the community. Progress in understanding the relationship between environmental and biological evolution in these communities, however, is hindered by the limited number of sedimentary sequences available for study and the difficulties in preserving evidence of biological development occurring at the intracellular level.

A major increase in the abundance of fossil microorganisms coincides with the growth of continents and the emergence of wide continental margins between 2.8 and 2.2 billion years ago. Microfossiliferous deposits are abundant in rocks between 2.5 and 0.6 billion years old (3,5). During this latter period, evidence of nucleated cells, multicelled life, formation of biogenic mineral deposits, and the persistence of atmospheric oxygen appears in the geological record. The availability of sedimentary sequences from this period offers opportunities for establishing causal relationships between the occurrence of glaciations, of oxygen in the atmosphere, and of periodic increases in geological activity (such as mountain building and growth of continents) and the manifestation of biological milestones, such as the advent of oxygenic photosynthesis, the development of planktonic eukaryotes and multicelled organisms, and the occurrence of episodes of evolutionary radiation.

Homologues of the ancient microbial ecosystems exist today in the form of microbial mats, the features of which are still controlled almost exclusively by unicellular life. Like the earliest ecosystems, these mat systems are also associated with hot springs and hydrothermal vents, as well as shallow hypersaline marine environments. The abundance, physiology, and ecology of the microorganisms in these contemporary systems should be studied as models for interpreting specific morphological, chemical, and isotopic features preserved in ancient rocks. Studies of both ancient and modern systems will be invaluable in establishing a knowledge base for carrying out the search for possible evidence of analogous biogenic structures on Mars.

In even the simplest of contemporary microorganisms, the complexity of the mechanisms for energy transduction, metabolism, replication, and translation argues for origins in much simpler apparatus. Vestiges of these primitive systems may still be preserved in the structures and functions of extant life. Although the biochemical and structural characterization of some enzyme systems has been investigated to determine the minimum requirements for retention of function, more work in this arena is needed, including efforts directed at ribosomal RNA's. Emphasis should be placed on understanding how the complex mechanisms found in organisms today may have developed from the simpler machinery. Studies from this evolutionary perspective hold promise of providing working models for the functional components of a minimal cell toward which research on prebiological chemical systems could be targeted.

A molecular phylogeny based on homologies in the nucleotide sequences of ribosomal RNA's has traced the ancestry of contemporary life back to three lines of descent from the primary kingdoms of eubacteria, archaebacteria, and eukaryotes.

Elucidating the evolutionary relationships of life on Earth using this approach depends largely on acquiring a broad data base. Toward this end, many more sequence data need to be acquired, especially among eukaryotic organisms and the uncultured organisms of all three kingdoms in hot spring, hydrothermal vent, and planktonic environments for which data are very sparse. Among the eukaryotes, the ultrastructural diversity reflecting phenotypic diversity should be studied in conjunction with the molecular studies. And integrated research on phylogenetic relationships and metabolic pathways of newly discovered organisms from all three kingdoms should be pursued to gain deeper insight into the evolution of metabolism. Thus, coupled molecular and phenotypic studies offer a quantitative means of determining the temporal sequence of early biological evolution, the chronology of which may be possible to establish with data from the geological record.

The earliest cellular organism must have been far simpler in terms of size and diversity of proteins, number and organization of genes, and biological specificity than any that exists today. Because the root of the universal phylogenetic tree has not been determined, however, the proximity of the universal ancestor of all life to any of the three kingdoms is unknown. Clues to the nature of this universal ancestor should be sought through comparative phylogenetic analyses of families of genes representing essential cellular functions among microorganisms of the three kingdoms. Common characteristics widely distributed in the earliest branching organisms in these kingdoms are expected to have been attributes of the universal ancestor. Identification of these traits should also yield insights into the geochemical and climatic conditions in the environment of the common ancestor.

Findings for Early Evolution of Life

- The geological record of biological evolution is lacking for the period spanning the earliest evolution of life on Earth prior to 3.5 billion years ago; it is sparse for the next billion years, and then increasingly accessible through the Precambrian.

- If the origin of life on Mars was contemporaneous with the rise of living systems on Earth, access to the earliest geological record of Martian sediments may yield the beginnings of a comparative paleontology among planets.

- Although paleontological and geochemical evidence exists in the geological record to relate the occurrence of biological radiations and innovations to environmental changes due to the physical evolution of the planet, these models need to be tested and refined against a broader data base.

- Remarkable progress is being made in establishing a universal phylogeny for life on Earth, and more can be expected as the extensive phylogenetic history preserved in organisms continues to be deciphered by means of molecular sequencing studies of their nucleic acids and phenotypic studies of their structures and functions.

Recommendations

- The Life Sciences and Solar System Exploration Divisions of OSSA should support the search for relatively unmetamorphosed Archean and Proterozoic sedimentary sequences; analogous samples from Mars should be acquired by unmanned missions and returned to Earth.

- The Exobiology Program should sponsor concerted studies of rock components of entire Precambrian sedimentary basins in which chemical, isotopic, paleontological, and paleoenvironmental information is simultaneously acquired on a common set of rocks.

- In the Exobiology Program, research on contemporary organisms aimed at unraveling the evolutionary history of life should focus on the following areas:

  — The abundance, physiology, and environment of the microorganisms in modern homologues of ancient microbial communities

  — Models of the simplest components of the apparatus required by microorganisms to carry out the indispensable energy harvesting, metabolic, and reproductive functions of life

  — The phylogeny and physiology of uncultivated organisms that inhabit hot springs, hydrothermal vents, and planktonic environments

  — The nature of the common ancestor of contemporary life as characterized by molecular phylogeny.

Evolution of Advanced Life

Research on the evolution of advanced life seeks to understand the influence of both intrinsic planetary processes and astrophysical processes on the course of biological evolution from unicellular to advanced forms of life. Such fundamental understanding will provide a basis for predicting the distribution of advanced life forms among other star systems throughout the galaxy. A direct search for technologically advanced life in the galaxy can be conducted by means of the Microwave Observing Project of the Search for Extraterrestrial Intelligence Program (6).

Studies of the evolution of metazoans and metaphytes during the most recent 600 million years show that the complexity of advanced life has not accumulated at a steady rate but rather episodically in surges that are rapid on a geological time scale (5). During this period, major environmental fluctuations occurred, including changes in the areas of shallow marine and continental habitats, regional climatic shifts, alteration of the geographic continuity of oceans and continents, onsets of glacial and thermal intervals, and variations in atmospheric and oceanic chemistries. These changes, due in part to the internal dynamics of the Earth, are expected to have influenced the course of evolution, but the causal relationships remain to be established.

Foremost among the possible extraterrestrial influences on biological evolution are the environmental perturbations resulting from impacts of asteroids and comets and the periodic cycles of climatic changes arising from the Milankovich cycles of orbital effects in the Earth-Moon-Sun system. The influence of the Milankovich cycles on climate over the last 500,000 years has been well documented, and evidence for such effects deeper in the past is being sought. The discovery of platinum group elemental anomalies in samples from global distributions of sediments deposited about 65 million years ago has led to the theory that impacts were responsible for mass extinctions at the Cretaceous-Tertiary Boundary. That these extinctions were followed by the rise of mammals to dominance underscores the need to assess the occurrence of such phenomena throughout the history of life. In addition to laboratory studies of the geological record, theoretical studies should be carried out to predict the perturbations of the terrestrial environment caused by large asteroidal impacts and the biological consequences of the resulting physical and chemical perturbations. Evidence of these predictions should be sought in the rocks that record the extinctions.

Understanding the relationship between biological evolution and the influences of endogenous planetary processes and exogenous astrophysical processes will require the gathering of information from diverse sources to construct a comprehensive paleontological data system. This system should incorporate detailed analysis of the fossil record of extinctions to the genus level; geological, geochemical, and paleoenvironmental data associated with that record; and corresponding data on the cratering record of impacts. With this data base, it should also be possible to assess the existence of periodicity in mass extinctions. Once found, the relationship, if any, can be determined between the periodicity and the records of extraterrestrial phenomena, such as impact craters on the Earth, Moon, or Mars; changes in the tidal interaction of the Moon and Earth; and variation of insolation resulting from cyclic and noncyclic changes in the orbit and axial inclination of the Earth.

Findings for Evolution of Advanced Life

- A new and exciting interdisciplinary field of science has emerged as it has become increasingly clear that, in addition to the effects of intrinsic geological activity, extraterrestrial phenomena due to Earth's cosmic environment may have played a critical role in influencing the course of biological evolution.

- The historical record of astrophysical phenomena, particularly asteroidal or cometary impacts, may be preserved in the rocks of the Moon and Mars.

Recommendations

- **The Life Sciences and Solar System Exploration Divisions of OSSA should increase support of research designed to determine the occurrence of elemental anomalies and other extraterrestrial signatures in the sedimentary record and their correlation with contemporaneous changes in the composition of fossil biota.**

- **NASA should include in its scientific objectives for future exploration of the Moon and Mars the search for evidence of impacts or other astrophysical phenomena that may be time correlated with analogous occurrences on Earth.**

Solar System Exploration

The goal of this component of the program is to determine the extent to which prebiological evolution has proceeded on other bodies in the solar system. This goal is accomplished by conducting experiments and analyses with spaceborne instruments to measure directly the elemental and chemical composition of comets, asteroids, and the atmospheres and surfaces of other planets and their satellites.

Exploration of the solar system has made possible the comparative study of planets (7,8). The knowledge gained indicates that even though some planets may form by similar processes from common building blocks and share a common early history, differences in size, location, composition, and other factors will eventually cause divergences in their subsequent development. For this reason, the prospects for the emergence of life on a planet are also inextricably tied to that planet's development. And data pertinent to the history and properties of planets and other objects in the solar system should be sought with spacecraft probes.

This montage of images was assembled from photographs of Saturn and its moons taken during the Voyager 1 *mission. Clockwise, the moons are Dione (in front of Saturn), Enceladus, Rhea, Titan, Mimas, and Tethys. A star background has been added by an artist.*

As a result of information returned from the *Viking* mission, the importance of seeking answers to questions about the nature of chemical evolution in ancient Martian environments, the possible origin and fate of life on Mars, and the relationship between the early histories of Mars and Earth has been strongly underscored. Missions to Mars in the next several decades include opportunities to address these issues with orbital observers and automated surface rovers and sample return. The Exobiology Program should be actively involved in all these missions. In these forays, the groundwork should be laid for conducting the future exploration of Mars by humans. Even issues as speculative as the feasibility of making Mars habitable for terrestrial organisms could be addressed by data provided by these opportunities (9).

The *Voyager* missions to the Outer Planets have stimulated exobiological interest in several of their satellites. Europa, the second major satellite of Jupiter, is covered with ice, but its size and density suggest that it may have a subsurface ocean of liquid water and may contain organic matter like that of carbonaceous meteorites. Titan, the satellite of Saturn, has an atmosphere containing nitrogen, methane, and a variety of simple organic compounds and is thought to represent a model of the chemically reduced atmosphere of the primitive Earth. The opportunity to conduct a detailed study of the organic chemistry of Titan's atmosphere should be pursued on the Titan/Cassini Mission.

Recent studies have revealed that many asteroids, Comet Halley, other comets, and cosmic dust particles thought to have been derived from comets contain an abundance of the biogenic elements in the form of organic matter. Insufficient data exist, however, to answer such intriguing questions as the degree of molecular complexity in this matter, the mechanisms for its synthesis, and its possible role in the origin and evolution of these primitive objects. Some answers may be provided by the direct study of comets and asteroids, as envisioned for the CRAF and Rosetta Missions, while others may be obtained by conducting experiments under microgravity conditions and by study of cosmic dust grains collected on the Space Station.

In order to participate in missions, it is necessary to design and construct highly sophisticated analytical instruments and experimental apparatus suitable for measuring the isotopic, chemical, and mineralogical composition of phases containing the biogenic elements on Mars, Titan, asteroids, and comets. These measurements will provide the basis for determining the origin of these phases and for assessing what relationships exist between the processes responsible for synthesis of extraterrestrial organic matter and those that produced the molecular precursors of living systems during prebiotic evolution on Earth.

Findings for Solar System Exploration

- Opportunities to fulfill the scientific objectives of Exobiology in space are provided in NASA plans for exploration of the solar system and construction of the Space Station.

- The exploitation of these opportunities will depend importantly on the development of the analytical flight instruments and apparatus needed to conduct research in space.

Recommendations

- **NASA should implement its plans for the following missions and facilities over the next several decades:**
  - Mars Rover/Sample Return Mission
  - Titan/Cassini Mission
  - Comet Rendezvous Asteroid Flyby Mission
  - Rosetta/Comet Nucleus Sample Return Mission
  - Space Station Cosmic Dust Collection Facility
  - Space Station Gas Grain Simulation Facility.

- The Exobiology Program should intensify efforts to fulfill its scientific objectives in space through participation in future missions and facilities.
- The Life Sciences Division of OSSA should increase its support of efforts to develop the advanced technology needed by the Exobiology Program to build instruments and apparatus for use in space.

Search for Extraterrestrial Intelligence

The Search for Extraterrestrial Intelligence is a research and development effort with the goal of determining the distribution of technologically competent civilizations in the galaxy. This goal is achieved by conducting a systematic search for artificially generated radio signals in the microwave portion of the electromagnetic spectrum.

Recent astronomical observations lend strong support to the astrophysical theory that planetary systems are commonly produced as a consequence of star formation. Current theories of chemical evolution and the origin of life predict that life will evolve on planets where the conditions are suitable. The development of intelligent life on Earth is perceived as an outgrowth of recent planetary evolution. Given the enormous number of stars, life may be very abundant in the galaxy. It is possible that intelligent life with technological civilizations may also be widespread.

In 1959, Giuseppe Cocconi and Philip Morrison proposed that radio transmissions in the neighborhood of the natural radio emanations of neutral hydrogen (1420 megahertz) might be a means by which civilizations could communicate over interstellar distances.

Following on the proposal by Cocconi and Morrison, the Life Sciences Division has conceived, developed, and demonstrated the technological capability to the point where it is ready to implement the most comprehensive search for extraterrestrial intelligence, the Microwave Observing Project (MOP) (6). Increasing radiofrequency interference, however, may pose a problem in the future.

The Microwave Observing Project has elements of strong public appeal, prospects for broad international cooperation, and unique scientific contributions to make to radioastronomy. The detection of extraterrestrial signals of intelligence would have profound impact on humankind.

Finding for SETI

- The readiness of the technology, the problems posed by increasing radiofrequency interference, and the strong public appeal of the SETI Program indicate that the time is ripe for implementation of the Microwave Observing Project.

Recommendation

- NASA should initiate the SETI Microwave Observing Project now and take the following steps toward its completion:
  - Build a fully functional, prototype SETI system, test it in the field, and use it to carry out the Microwave Observing Project.
  - Conduct the Targeted Search and Sky Survey of the SETI MOP.

Relationships Between Exobiology and Other Research Programs

No basic research program comparable in scientific scope to the Exobiology Program exists elsewhere in the world. The goals and objectives of the program are of great interest among scientists around the world, and the number of investigators conducting exobiological research in Europe, Japan, India, and the Soviet Union is growing. NASA should encourage the development of a broad international community of exobiologists to stimulate the research area and to expand awareness of the unique contributions made by its space missions to this fundamental field of research.

Unlike the discipline-oriented scientific investigations supported by the National Science Foundation (NSF) and other funding agencies, the research encompassed by the Exobiology Program is strongly interdisciplinary and often mission oriented. Studies of Antarctic microbial ecosystems, however, are supported jointly by NASA's Exobiology Program and by the NSF's Division of Polar Programs. The opportunity to broaden the active scientific constituency of the Exobiology Program suggests that more such coordination would be beneficial.

The Exobiology Program is closely connected to other NASA programs. Because it deals in large measure with the history of life on Earth, the Exobiology Program establishes an interface with the Life Sciences Division's Biospherics Research Program, which is concerned with understanding the present relationship between life and its environment on Earth. In principle, hypotheses generated by the Biospherics Research Program, as well as related programs in OSSA's Earth Science and Applications Division, can be tested through study of the geological record. In turn, interpretations of the geological record of biological evolution can be assessed in light of knowledge of the present biosphere-geosphere system. For these reasons, the Life Sciences Division should exploit opportunities to coordinate activities of the Exobiology and the Biospherics Research Programs that will lead to mutual enhancement.

The scientific goals of the Exobiology Program also complement those of other divisions within NASA. Investigations of other bodies in the solar system for information pertinent to the origin of life or its precursors formed an integral part of the science objectives identified by the Solar System Exploration Committee of the NASA Advisory Council in its reports, *Planetary Exploration Through Year 2000: Part One: A Core Program* (7), and *Planetary Exploration Through Year 2000: Part Two:*

An Augmented Program (8). Similarly, the National Academy of Sciences' (NAS) report of the Astronomy Survey Committee, *Astronomy and Astrophysics for the 1980's* (10), recognized the scientific value of astronomical studies of the organic chemistry of the galaxy and the existence of extrasolar planets, and it recommended that the SETI be initiated.

Findings for Relationships Between Exobiology and Other Research Programs

- The Exobiology Program is unique to NASA, and it has broad public appeal.

- Sound scientific interrelationships between the Exobiology Program and other research programs in OSSA form a strong basis for ongoing cooperation between the Life Sciences, Solar System Exploration, and Astrophysics Divisions.

- Space missions conducted by the Solar System Exploration Division are essential for carrying out exobiological investigations into the rest of the solar system. The great observatories in space provided by the Astronomy and Astrophysics Division afford platforms for astronomical studies of the nature, abundance, and distribution of biogenic elements and compounds throughout the galaxy.

Recommendations

- **OSSA should foster the unique character of the Exobiology Program by supporting its scientific goals and objectives across the Agency and continuing to provide a balance of opportunities to conduct mission-oriented and ground-based fundamental research within NASA.**

- **OSSA should develop cooperative plans for using space missions to pursue scientific objectives pertinent to Exobiology whenever interests in the objectives are shared among the Life Sciences, Solar System Exploration, and Astrophysics Divisions of OSSA.**

- **The Life Sciences Division should expand coordination between the NASA Exobiology Program and related NSF programs to explore areas of mutual scientific interest that may prove fruitful to pursue cooperatively.**

Program Management and Administration

The Exobiology Program addresses fundamental questions about the origin and evolution of life and intelligence that can be controversial in nature and profound in their significance. Therefore, it is necessary to uphold the high scientific quality and credibility of the program by maintaining high standards of excellence and scientific rigor in the peer-review process for all funded research proposals.

Historically in the Exobiology Program, the concepts for experiments and measurements to be made on projects or space missions originated in the ground-based research. In that milieu, measurement requirements are defined and the concepts are tested for feasibility. The next phase involves the advanced development of technology necessary to build the hardware. The final stage is implementation of the mission or project. These stages are represented by three functional elements in the current structure of the Exobiology Program: 1) ground-based research, 2) pre-project advanced development, and 3) missions and projects. The present program organization of functional components and evolutionary epochs should be maintained.

The scope of the Exobiology Program embraces an extremely broad range of technical expertise and scientific disciplines. Because many of the scientific problems in the Exobiology Program lie at the interface between scientific disciplines, they are most effectively addressed with a multidisciplinary approach. This is particularly true for field- and mission-oriented research and studies of small or rare samples of biological or geological origin on which many correlated measurements must be made. The disciplines involved may be as seemingly disparate as astrophysics and biochemistry, or as related as organic chemistry and biology. Often, practitioners of the separate disciplines are unaware of the contributions each can make to the overall effort. The integration of such diverse research elements is critically important. Toward this end, the Exobiology Program should conduct regular, topical, multidisciplinary science workshops in which key scientists representing all areas germane to the topic are assembled to share and synthesize knowledge, identify fruitful research approaches and future directions, and develop collaborative activities.

Findings for Program Management and Administration

- A vigorous Exobiology research program requires the participation of scientists from many disciplines in NASA and the scientific community at large, some of whom may already be associated with research programs of other divisions or other agencies.

- The many years usually required to address adequately a major research problem on the ground or to translate a ground-based research effort into an experiment or project in space underscores the need for long-term commitments on the parts of both the Exobiology Program and many of its investigators.

- Most academic institutions train young scientists in strongly discipline-oriented departments where exposure to the science of exobiology may be minimal. Yet young scientists capable of conducting interdisciplinary research are the lifeblood of the Exobiology Program.

Recommendations

- **NASA should maintain a strong multidisciplinary team of inhouse scientists with the technical expertise and programmatic commitment to assist the program manager in developing future programs and an external scientific constituency.**

- **The Exobiology Program should institute a policy to support at any given time at least one multidisciplinary team of investigators selected by peer** review to address an opportune research effort that is long term and ground based.

- **NASA should establish broad channels of communication between disciplines and across internal organizational boundaries and between the Agency and the scientific community at large. In addition, it should ensure wide dissemination of announcements about opportunities to conduct research, to develop new program thrusts, and to participate in summer study, graduate research, and postdoctoral research programs.**

- **The Life Sciences Division should cultivate the interests of young scientists in exobiology through increased support of student internships, summer study programs, graduate research programs, and postdoctoral fellowships to be held at both universities and NASA research centers.**

Funding

Funding for the Exobiology Program over the past 4 years has been maintained at a relatively constant level of about $6 million, which corresponds to about 9 percent of the Life Sciences Division's budget. During this interval, two program elements were added — Cosmic Evolution of the Biogenic Compounds and the Evolution of Advanced Life — to complete the evolutionary scope of the program. At the same time, the program was committed to addressing an increasing need for access to space missions and for initiation of a project activity. As a result, fewer resources are available for basic research in exobiology.

If funding continues at a relatively constant level, the fundamental ground-based research program will either continue to be eroded or it will be forced to retrench by reducing its intellectual scope and, therefore, its excitement and challenge. Only substantial additional funding would allow the program to overcome losses due to inflation and the ever-increasing rise in overhead. It would then be in a strong position to capitalize on mission opportunities that NASA so uniquely provides, and NASA would retain its mantle of leadership in this scientific arena.

Finding for Funding

- An essentially constant level of funding during a period of increasing demand for development of project activities has spread resources very thinly and seriously eroded the ground-based research program without providing adequate stimulus to the project activities.

Recommendations

- **NASA should significantly enhance the ground- and space-based research capabilities and infrastructure (funding, inhouse manpower, and facilities) of the Exobiology Program in order to maintain the Agency's leadership role, to implement the science strategies recommended by NASA and NAS advisory committees, to capitalize on the existing data base, and to optimize the design and scientific return of future missions.**

- NASA should increase support for the technology development program necessary to generate advanced systems for instrumental analyses, remote sensing, and data analysis for use in future missions, particularly those systems that will be essential to optimization of science returns from Mars exploration programs.

Reference List

1. National Academy of Sciences. Committee on Planetary Biology and Chemical Evolution. In press, 1988. "Planetary Biology and Chemical Evolution: Progress and Future Directions." Washington, DC: National Academy Press.

2. Wood, J.A., and S. Chang, eds. 1985. *The Cosmic History of the Biogenic Elements and Compounds*. NASA SP-476. Washington, DC: National Aeronautics and Space Administration.

3. Hartman, H., J.G. Lawless, and P. Morrison, eds. 1985. *Search for the Universal Ancestor*. NASA SP-477. Washington, DC: U.S. Government Printing Office.

4. DeFrees, J.G., D. Brownlee, J. Tarter, D.A. Usher, W.M. Irvine, and H.P. Klein. In press, 1988. "Exobiology in Earth Orbit." Washington, DC: National Aeronautics and Space Administration.

5. Milne, D., D. Raup, J. Billingham, K. Niklaus, and K. Padian. 1985. *The Evolution of Complex and Higher Organisms*. NASA SP-478. Washington, DC: National Aeronautics and Space Administration.

6. Drake, F., J.H. Wolfe, and C.L. Seeger, eds. 1983. *SETI Science Working Group Report*. NASA Technical Paper 2244. Washington, DC: National Aeronautics and Space Administration.

7. National Aeronautics and Space Administration Advisory Council. Solar System Exploration Committee. 1983. *Planetary Exploration Through Year 2000: Part One: A Core Program*. Washington, DC: U.S. Government Printing Office.

8. National Aeronautics and Space Administration Advisory Council. Solar System Exploration Committee. 1986. *Planetary Exploration Through Year 2000: Part Two: An Augmented Program*. Washington, DC: U.S. Government Printing Office.

9. National Commission on Space. May 1986. *Pioneering the Space Frontier*. New York: Bantam Books.

10. National Academy of Sciences. Astronomy Survey Committee. 1982. *Astronomy and Astrophysics for the 1980's. Vol. 1: Report of the Astronomy Survey Committee*. Washington, DC: National Academy Press.

William C. Schneider, D.Sci.
Chairperson

Gerald P. Carr, P.E., D.Sci.

Michael Collins

Keith Cowing
Staff Associate

Mitchell K. Hobish, Ph.D.
Staff Associate

Lauren Leveton, Ph.D.
Staff Associate

Flight Programs

Summary

The NASA Life Sciences Division has a well-developed and well-understood set of strategic objectives for the 1990's: extending crew stay time for the Space Station to at least 180 days; understanding the human physiological and psychological requirements for long-duration missions to other planets, such as a round trip to Mars; understanding the physiological and psychological needs for extended living in gravitational fields of less than 1 Earth gravity (g); and using the unique environments of space to better understand biological and physiological processes in 1 g.

Each of these objectives has emphases common to the other objectives. Each would benefit by complementary and supportive elements conducted in a number of flight projects. Each could be furthered by a program plan that calls for diversified flight opportunities: short-duration, human-tended projects, such as the Shuttle, Spacelab, and Spacehab; longer duration experiments flown on recoverable flight projects, including Lifesat, Biocosmos, the Commercially Developed Space Facility (CDSF), and the Space Pallet Satellite (SPAS); and long-duration experiments using the Space Station as a base. Maximum progress can be made at minimum cost if flight experiments are selected that support multiple strategic objectives and a variety of flight projects.

Introduction: NASA's Mandate and the Life Sciences

NASA is a mission-oriented organization dedicated to conducting the research and engineering necessary to explore space. The Agency was chartered to contribute to "the expansion of human knowledge of phenomena in the atmosphere and space" and to "develop vehicles capable of carrying instruments, equipment, supplies and living organisms through space" (NASA Act of 1958, Section 102 [c][1][4]). While the relationship between scientific research and engineering is synergistic, it requires cultivation and depends on close coordination among various disciplines.

NASA conducts life sciences research for two reasons: to understand basic biological processes and to support the presence of humans in space. These two efforts are often intertwined, with a finding in one area often leading to a

The STS 51-B Spacelab mission begins with the liftoff of orbiter Challenger *from Pad A at 12:02 p.m. on April 29, 1985.*

discovery in the other. Some significant questions in biology, addressing both basic biological processes and human health, can be answered only by space-based experimentation.

NASA's plans call for a greatly increased human role in space. Programs such as the Space Station and proposals for lunar and Martian exploration will require extensive basic and applied research. They will also require the Agency to assess and build on past efforts and accomplishments.

Personnel

Science is advanced by having many curious and innovative individuals generating ideas and experiments to test the ideas. Opportunities must be available, however, for the implementation of those experiments. Very real problems exist in this area when it comes to space and space life sciences research.

Investigators who have already been promised flight time see previous flight delays compounded even further. Graduate students, laboratory space, and institutional support become more difficult to justify. People who had been considering the submission of a proposed experiment see the wait time as being better spent pursuing projects that can promise a faster turnaround. This is especially unfortunate when young graduate students want to pursue a career in space research but see that their proposal might not fly until several years after the date they had planned to finish their education.

Space life sciences, if they are to attract and hold those minds most able to use the space environment to discover the secrets of nature, must offer participants the opportunity to experiment on a regular and frequent basis.

Findings

- The health of any scientific discipline is directly related to the availability of individuals to perform needed research. Ensuring that such a community is accessible starts with education. NASA has always made great efforts to involve students in its activities. NASA's Life Sciences Division has made special attempts to involve students directly in actual research. Excellent examples include the Space Life Sciences Training Program oriented toward undergraduates and a number of graduate assistantship programs.

- Significant efforts have already been made toward the establishment of formal working relationships with other Government agencies and research organizations, such as the National Institutes of Health, the National Science Foundation, the National Research Council, the National Oceanic and Atmospheric Administration, and the Department of Defense.

Recommendations

- **NASA should greatly expand its efforts to attract and support new space life sciences researchers. A special effort should be made to support the students who will become the scientific investigators of the 21st century.**

- Researchers outside the space life sciences community should be actively encouraged to participate in space-related research. Formal ties to governmental agencies and research organizations should continue to be established and strengthened.

Performance of Space Life Sciences Research: Access to Space

The only way to observe the effects of space flight on living organisms is to fly appropriate research specimens aboard spacecraft. Biological processes are extraordinarily dynamic and involve the interaction of many internal and external environmental factors. Because biological research is largely experimental, statistically significant sample sizes must be exposed to environmental variables. On-orbit controls to isolate the effects of different variables are mandatory. Various exposure times, orbital inclinations, and reflight opportunities must also be available to investigators.

Data obtained from flight experiments must be accessible to investigators in a timely manner for them to analyze results, refine models, and build upon previous knowledge. It should be emphasized that ground and flight programs are part of a continuum and cannot be separated from one another. A schematic description follows of a research paradigm that shows the interdependence of all aspects of a solid space life sciences research program:

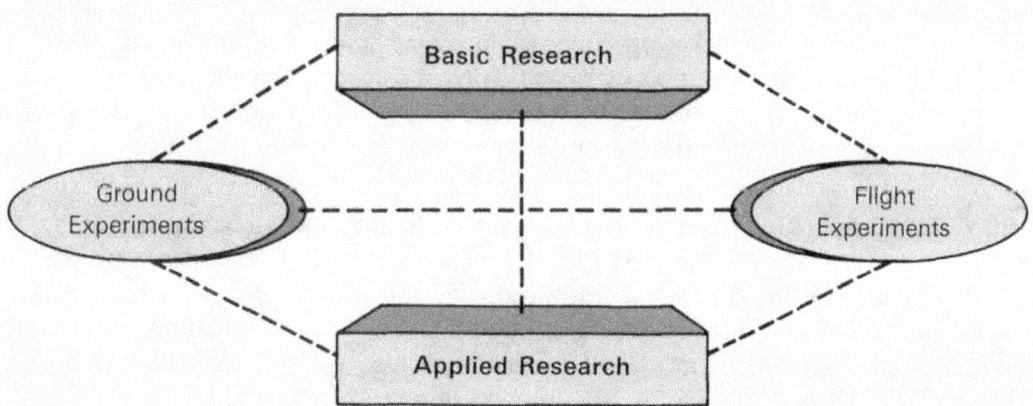

Nearly all space-based life sciences research currently funded by NASA is designed to be performed aboard the Space Transportation System (STS), which has not flown since the *Challenger* accident in January 1986. Once the STS resumes operations, payload space will be at a premium. In the coming years, it will be largely dedicated to Space Station assembly and operations and Department of Defense missions.

To ensure safe operation of the Space Station, substantial life sciences research must be performed. NASA investigations are currently limited to analysis of earlier space flight data and information derived from ongoing ground-based work. Except for some experiments conducted by American investigators aboard the

Soviet Biocosmos satellite missions, NASA cannot currently conduct life sciences research in space. As presently envisioned, planned resources will not support all required life sciences research. Therefore, an alternate means of gaining access to space is mandatory.

In the aftermath of the *Challenger* accident, NASA conducted a study aimed at developing a mixed fleet of launch vehicles. Out of this study came the impetus for the Life Sciences Division's proposal for a free-flying satellite dedicated solely to life sciences research. Such a vehicle has proved a useful research tool in the past: NASA flew three similar satellites during the Biosatellite Program in the 1960's. While the first satellite was lost during the recovery phase, the other two missions were successful and yielded important data. The Soviet Union has been flying its own version of this concept, the Biocosmos series (which is based upon a modified *Vostok* spacecraft), for over a decade with similar success. American participation in the Soviet program has provided significant information on the effects of space flight on the musculoskeletal system and on radiation effects at high inclinations.

As currently envisioned, this new satellite program would use an expendable launch vehicle (ELV) with an autonomous return capability very similar to that used for Biosatellite and Biocosmos. Such an autonomous system would offer a number of capabilities unavailable or unfeasible with the STS: a flexible, independent launch schedule; mission durations of 30 days or more; unique orbital altitudes and higher orbital inclinations, including polar orbits (of special interest in determining radiation effects); simplified and standardized hardware design; and rapid turnaround, with two or more flights per year. This system also affords the possibility of international participation: Several other spacefaring nations and international space agencies have expressed interest in this concept.

Since flight opportunities and payload space will continue to be limited, it is imperative that NASA make best use of available resources. Significant preparation must be done on the ground, including the design, development, testing, and evaluation of equipment; development and testing of experimental protocols; and computer simulations. Ground-based research is also essential to develop models that replicate all or some of the phenomena observed in space.

Many opportunities to conduct experiments are possible if NASA gives sufficient priority to the life sciences. Experiments may be accommodated on Shuttle middeck lockers and on Spacelab, on international missions, such as Biocosmos and the reusable SPAS (West Germany), and on future missions, such as Spacehab and the Commercially Developed Space Facility. As previously noted, a free-flying life sciences satellite is also a realizable asset.

Findings
- A large backlog of approved life sciences experiments has yet to fly. The time between announcement, selection, and flight can exceed a decade. These

delays wreak havoc with the research programs of individual scientists and ultimately work to the detriment of the entire space life sciences community.

- Many experiments require prompt reflight to increase sample size, to validate experimental design and hardware, and to rerun inconclusive or malfunctioning investigations. This capability has rarely been available to life scientists.
- A variety of alternative means of gaining access to space already exists or could be developed.
- The United States would benefit greatly in a number of areas by cooperating with foreign space programs. While progress has been made in this regard, much more needs to be done.

Recommendations

- NASA should do the following to reduce the delay between the Agency's acceptance of a proposal for a flight experiment and actual launch of that experiment:
  — Continue the establishment of Discipline Working Groups, which allow greater contact between investigators and NASA programs when experiments are solicited.
  — Limit the scope of Announcements of Opportunity by making them more discipline-oriented.
  — Try to establish a firmer link between Announcements of Opportunity and specific, manifested missions.
  — Link Announcements of Opportunity with theme-oriented missions or programs whenever possible.
  — Target different Announcements of Opportunity to different experimental opportunities available on the Space Shuttle middeck, Spacelab, free-fliers, Space Station, Biocosmos, and elsewhere.
  — Release Announcements of Opportunity on a regular basis to allow potential researchers to plan their proposal preparation and resources better.
  — Accept a smaller number of experiments with more narrowly targeted objectives to prevent overlap and maximize resources.
- Life sciences payloads should be given priority so that life sciences research is routinely conducted in space.
  — Payload space, such as Shuttle middeck lockers, should be made available on a priority basis on each Shuttle mission for life sciences research.
  — If middeck lockers are not available, NASA should look into the availability of Spacehab, the Commercially Developed Space Facility, or other facilities as substitutes.

- An extended-duration Shuttle orbiter should be considered a useful resource, especially to test technology modifications (such as exercise equipment) for the Space Station's Health Maintenance Facility.

• NASA should develop a recoverable life sciences spacecraft equipped with variable-gravity capabilities that can be used for animal (rhesus monkey-sized), plant, and cellular research.

• A systematic approach to cooperation and coordination with other spacefaring nations should be strengthened.

- NASA should continue its participation in established working groups and explore ways in which this collaboration can be expanded.

- The Soviet Union and the European Space Agency have important programs, as do the German Research and Development Institute for Air and Space Travel, the French National Center for Space Studies, the National Space Development Agency of Japan, and other national space agencies.

• Collaborative efforts with the Soviets should be vigorously pursued and expanded.

- This should be done in part by participation on future Biocosmos missions. Participation in this program has provided data useful in understanding the effects of space flight on humans.

- The possibility of flying experiments aboard the Soviet Space Station and cooperation in planning between the Biocosmos and NASA free-flying satellite efforts should also be explored. Such cooperation was initiated several years ago, especially with joint data analysis (musculoskeletal for pre- and postflight missions).

- Opportunities for reciprocation by the U.S.S.R. on U.S. missions should be explored.

• NASA should lead in the development of a standardized international biomedical data base that will allow all spacefaring nations to share information. An automated and easily accessible data base is needed today, just to provide NASA with the currently available archival data on a routine basis.

The Space Environment: Potential Limitations on Extended Human Presence in Space

As is the case with all science disciplines, the life sciences community is concerned with developing a basic understanding of the world around us. Life sciences, particularly the medical sciences, are also applications driven. Life sciences investigations are unlike those pursued by sister sciences in that it is rare that a rigid formula or natural equation can be derived (except, of course, for chemical equations). Life sciences are, in general, empirical by nature and characterized by a statistical analysis of observed relationships. Since life sciences investigations deal with complex and incompletely understood interactions, repetitive and rapidly executed variations of experiments and a large statistical base are essential for understanding basic life processes. Moreover, investigators are needed to monitor experiments conducted in a space laboratory and to analyze research results.

Long-duration human missions pose considerable challenges for life scientists. The times required for round-trip journeys to other planets are measured in years. Since no analog on Earth duplicates the environment of space, the investigation of long-term human tolerance to space flight requires the use of facilities such as the Space Station. Tests of human competence to withstand years of space travel must be initiated and validated long before human interplanetary missions are launched. Not to do so would risk significant delays or even cancellation of such missions.

Findings

- The effects of zero or partial gravity on humans and other forms of life are not fully understood.

- The radiation environment in space, particularly galactic radiation, is not fully understood.
  - Standards for exposure to galactic radiation have not been established with any degree of confidence.
  - Research into the effects of radiation exposure in space will necessitate orbital inclinations and periods that are often unobtainable or impractical with the STS.

- Current mission planning requires extended human space flight. American space-flight experience is limited to 84 days. While the Soviet Union has collected data on one individual from a 326-day flight, its information on human space flight exceeds ours only with regard to cumulative number of days spent in space. The Soviet data are mostly of an observational or operational nature. Extended-duration missions, such as the exploration of Mars, will require stay times of a year or more.
  - Available life sciences data are insufficient to support the design of a system that would use centrifugal force as a long-term substitute for gravity.
  - In addition, data are insufficient and incomplete to support the use of proposed countermeasures to alleviate problems associated with long-duration space missions.

- Research has to be expanded to understand the reaction of living organisms to the space environment.
  - While mission planning will revolve around the specific physiology of humans, substantial amounts of applicable data can be derived from experiments on nonhuman subjects.
  - A vigorous program of animal research is vital to extrapolate the effects of

space flight to humans and to understand the effects of countermeasures, including the use of artificial gravity.

- Spacecraft intended for extended space missions cannot be designed to support human crews safely if the physiological and environmental specifications have not been reliably defined. It would be inadvisable to commit to either a weightless or an artificial gravity-based vehicle design before the basic reactions of humans to space flight are fully understood.

- Without appropriate life sciences research and technology programs and dedicated spacecraft, inadequate subsystem design and operational scheduling would likely lead to increased costs resulting from higher frequency of Shuttle logistics flights, lower overall crew productivity, and decreased mission effectiveness.

- NASA plans call for the Space Station to serve as primary test facility for long-duration space research. Life sciences and microgravity science have been identified as the two major users of the Space Station.
    - The specific accommodations for life sciences research on the Space Station remain vague.
    - This uncertainty is problematic to both the life sciences and microgravity science communities inasmuch as concerns have been raised that Space Station research in these areas could be mutually incompatible and that the projected research has not been properly addressed from the systems engineering point of view.

- The Space Station will not be appropriate for all life sciences research. In addition, it may not be available for all the research that needs to be performed. Many experiments, such as radiation or artificial gravity research, will need to fly aboard spacecraft with mission characteristics that are impossible or unfeasible to achieve with current plans for the Shuttle and Space Station.

Recommendations

- Ground-based programs must be expanded to support research aimed at extending the human presence in space.
    - Ground-based research will help develop countermeasures to prevent or accommodate space adaptation during long-duration space flight. Mission options using microgravity and countermeasures versus artificial gravity should be researched in parallel. Results from both efforts should be analyzed and used to modify one another as the feasibility and efficacy of each becomes apparent. This will result in shared resources, cost reduction, and a decrease in the amount of lead time needed to build hardware for long-duration missions.
    - Ground-based studies can help identify, quantify, and resolve human factors limitations.

- In addition, such research can identify, quantify, and develop countermeasures to radiation hazards likely to confront humans during extended-duration missions.

- Timely phasing of research and technology (R&T) activities in the life sciences is an absolute requirement.
  - Such activities must begin immediately, since the results of R&T will have long-reaching effects.
  - Appropriate ground-based facilities are required to support life sciences R&T.
  - A long-duration, free-flying bioplatform capability is needed to conduct life sciences research for periods longer than 20 days.

- The Space Station should be furnished with research facilities and instruments to support experiments leading to stay times of up to 2 years as an analog for human missions to Mars. A dedicated laboratory for life sciences research must be provided. It should be designed to allow evaluation of potential Controlled Ecological Life Support Systems (CELSS) designs and simulation of the isolation a crew might experience on long-duration missions. In addition, it should have the capability of isolation in case of environmental contamination.

- The Space Station and Spacelab should both be furnished with a variable-gravity facility that includes a 1.8-meter centrifuge rated for performing small animal and plant investigations. The use of larger diameter centrifuge facilities for human studies should be thoroughly studied.

- Devices designed to measure and record remotely all aspects of ambient radiation environments, specifically galactic radiation, should be placed on all high-orbit and interplanetary spacecraft. This will require additional resources for the development of better instrumentation, including real-time telemetry and data acquisition.

Instrumentation and Computational Capabilities

Taking an experiment from the concept stage to flight is a complicated and time-consuming process. The rules governing experimental hardware design have always been dictated by the unique restrictions imposed by spacecraft design and operations. The challenges of the design and development process, coupled with a limited number of flight opportunities, cause NASA to place a premium on deriving the most scientific value out of every experiment. Historically, NASA has flown each experiment once, with no guarantee of reflight or of follow-on experiments. To ensure that flight equipment would return meaningful data with a high level of confidence, the Agency has often limited the scope of an experiment to what can reasonably be performed in space. The process has often been success driven, the logic being that if NASA could only provide one opportunity for an experiment, the investigator(s) could only ask a question that the experimental hardware had a good chance of answering. As a result, hardware has often been custom tailored to perform one well-defined experiment on one flight without contingency planning for reflight. While the equipment was designed well for its one mission, it often could not be modified easily for use in other experiments because of resource and time limitations.

This situation changed dramatically with the advent of Space Shuttle operations and the development of a large number of versatile and reusable hardware items collectively known as Life Sciences Laboratory Equipment (LSLE). Often, an experiment will call for the development of hardware not already available. In such a situation, Experiment Unique Hardware (EUE) is developed by NASA, flown, and then added to the standing LSLE stock for use by subsequent investigators.

The Space Shuttle will never fly as often, nor will it be as flexible or inexpensive, as its planners hoped it would be. In the aftermath of the *Challenger* accident, the limited number of flight opportunities was reduced even further. Therefore, each flight opportunity must be fully exploited by standardizing experimental equipment and protocols, making the best use of personnel resources, and avoiding programmatic overlaps.

Findings

- Crew members are generally willing to participate as test subjects if they are informed of the scope and significance of research objectives and if they can expect to benefit from these objectives. There is, however, significant resistance to such participation if it involves the use of invasive probes. Budgetary limitations have precluded the development of appropriate, noninvasive, state-of-the-art instrumentation and have forced the use of off-the-shelf equipment, which sometimes means invasive instrumentation.

- While NASA strives to maintain a position near the leading edge of advanced technology, the state-of-the-art in instrumentation and in computer design and architecture is changing more quickly than the Agency can accommodate.

- Instrumentation and techniques used by non-space life sciences researchers and space life sciences researchers are not always compatible. Extrapolation from findings obtained in one area to the other is not always possible.

Recommendations

- **Noninvasive monitoring instrumentation should be developed to provide physiological data equivalent to that obtained with more traditional, invasive techniques.**
- **NASA should invest suitable resources to ensure that the computational capabilities available for life sciences research are commensurate with the evolving state-of-the-art.**
- **Greater efforts should be made to reuse flight hardware.**
  - The LSLE hardware collection should be increased.
  - Engineering models and space-based experimentation protocols should be made more easily available to life sciences investigators.

Space Flight and Multidisciplinary Research

Some of the greatest technological and programmatic challenges confronting NASA's Life Sciences program are found in the most multidisciplinary research areas, addressed by Exobiology, Biospherics Research, and CELSS. These challenges are both scientific and organizational: the three areas are inherently multidisciplinary, with research objectives that span virtually the entire suite of activities sponsored by the Life Sciences Division, ranging from nucleosynthesis of biogenic elements, to molecular genetics, to atmospheric physics.

Such broad programs can achieve significant progress only through the synthesis of data, insights, and developments in the disciplines of biology, chemistry, climatology, computer sciences, engineering, geology, physics, and more. The data, insights, and progress are derived from a blend of ground-based laboratory, field, observational, and theoretical research, as well as from information collected by spaceborne laboratories, solar system exploration missions, and orbiting observatories.

The Exobiology Program has established tight interfaces with several other NASA programs, and it supports work conducted in cooperation with the National Science Foundation. As a result, Exobiology has developed techniques to facilitate the transfer of scientific, programmatic, and organizational information across the involved disciplines. Because it deals with fundamental questions that are often controversial, maintenance of rigorous scientific excellence and credibility is of paramount importance. Consequently, the guiding management philosophy of the Exobiology Program has been to create and maintain, by policy and administration, a climate that promotes communication across disciplinary and organizational boundaries, that fosters creativity and the development and implementation of new concepts, and that contains sufficient controls to assure scientific excellence. As such, the program may provide a model for NASA in implementing cross-disciplinary transfers for other aspects of its mandated activities in general and for the Life Sciences Division in particular.

Findings

- In support of Exobiology, Biospherics Research, and CELSS, studies of the chemistry of terrestrial and extraterrestrial environments would provide technical data directly applicable to designing experiments and instrumentation for solar system exploration, including investigations of planet Earth, and establishing requirements for support of long-term human space flight.

    — Understanding the relationship between biological evolution and the evolution of the Earth would assist studies of planetary and biological evolution elsewhere in the universe.

    — Flight missions are required to collect such data.

- The availability of capabilities for remote automated analyses of samples of planetary atmospheres and surfaces, without return to Earth, would enhance the attainment of scientific objectives for planetary biology research programs and allow assessments of the potential of surface sites for future habitability.

- From the standpoints of both basic and applied research, Exobiology, Biospherics Research, and CELSS fit well within the context of NASA's charter. They variously make use of NASA's unique capabilities for exploring bodies of the solar system (including, most emphatically, planet Earth) and observing astrophysical objects and events.

Recommendations

- **The transfer of information and technology among NASA divisions and between NASA divisions and other research organizations should be facilitated whenever possible.**

- **Efforts should be accelerated to develop devices for remotely collecting and analyzing samples of planetary atmospheres and surfaces, as well as remotely acquiring, storing, and analyzing planetary biological data. The objectives should be to characterize environments at remote sites without requiring sample return and to ensure that any limited samples that are returned are the most interesting scientifically.**

Strategic Approach

Clearly, biomedical investigation leading to an understanding of how and why the human body reacts to space flight must continue. The primary objective of this research is to have sufficient understanding of and control over the phenomena so that a single 2-year experiment with human subjects on the Space Station has a high probability of success.

One research approach would be to develop ground-based experiments using animals and computer simulations to identify human responses. Such an approach, along with short-term flights, can provide basic insights into the effects of extended human space flight and may lead to the development of appropriate countermeasures. A "proof of concept," full-duration mission on the Space Station is, however, mandatory. Such a test will require an isolatable, independent module to ensure that the test subjects and the test objectives are not compromised by contact with the transient crew members.

Findings

- A program requiring 180-day stay times aboard the Space Station is under serious consideration. Other programs that would place humans on the Moon and, eventually, on Mars are also being evaluated.
    - Implementation of such programs will entail the design and development of substantial amounts of experimental hardware and complex technologies.

- Space vehicles cannot, however, be designed to respond to human requirements when specifications for such requirements do not yet exist. Such specifications can only be advanced after basic biological research requirements have been defined.

- Research being conducted by life sciences at NASA has far-reaching consequences, not only in answering basic questions, but in supporting practical projects, such as determining the effects of gravity on CELSS and the concomitant ability to support long-duration, human space flight.

Recommendations for NASA

- Expand the use of space probes and ground-based techniques to examine the physical and chemical characteristics of planets and other bodies.

- Use a variety of manned and unmanned spacecraft, as well as ground-based facilities, to study the effects of different aspects of the space environment upon living systems.

- Use the Space Station to conduct a research program that will result in the ability to support humans safely and productively in space for periods up to 180 days and beyond.

- Focus efforts on developing a more fundamental understanding of the biological processes that limit humans in space and identifying appropriate countermeasures by:

  - Building up knowledge through a systematic, step-by-step approach examining a wide variety of concepts prior to embarking on full-fledged examinations.

  - Conducting a series of experiments on analog systems (biological or computational) to determine fundamental mechanisms.

  - Using all flight opportunities available to understand the major limitations to long-duration human space flight early enough so that appropriate preventive measures are tested and validated. It should also be kept in mind that total control of risk might not be practical or feasible.

  - Using the Space Station to prove concepts and countermeasures.

- Reevaluate the procedures for developing and selecting flight experiments to ensure timely research. The requirements for the experiments must meet certain criteria, such as an interval of success probability, timeliness, accommodation of existing technologies, crew scheduling and training, ease/cost of implementation, payoff or impact (short- or long-term), and collaboration with others.

- Establish working relationships and working groups among NASA, the NIH, and other research institutions and industries to develop a mutual understanding of fundamental biological processes and of measures to manage potential limitations to our conquest of the space environment.

- Increase the support of life sciences research projects at universities and other research institutions. A special emphasis should be placed on involving students in space research. The objective is to expand the base of life scientists participating in NASA programs and to assure that America can retain its competitiveness in space research.

- Enhance the accessibility of space to life sciences researchers by increasing flight opportunities and broadening the base of the Agency's contact with the entire spectrum of life sciences research.

Conclusions

NASA should support a vigorous program of flight projects to address strategic objectives. Specifically, it should:

- Develop a recoverable, reusable space platform that has a variable-gravity facility, can support a variety of flight experiments, and is designed for rapid turnaround. This capsule should be launched by a reliable, expendable vehicle.

- Allocate a greater number of Shuttle middeck lockers to life sciences experimentation and/or explore the use of Spacelab for that purpose.

- Increase the flight rate (priority) of Spacelab and dedicate a larger percentage of space, time, and resources to life sciences issues.

- Recognize the vital importance of the Space Station to the strategic objectives of the life sciences and allocate sufficient Space Station resources to those ends. Design the Space Station to include laboratories for clinical and biological research.

- Develop instrumentation for the remote determination of the environment, particularly cosmic radiation, and place those instruments on all appropriate spacecraft, especially geosynchronous and interplanetary.

- Develop instrumentation for noninvasive monitoring of the physiological status of subjects with an accuracy at least equal to that available with current invasive techniques.

Reference List

Alexander, Joseph K., Philip C. Johnson, Percival D. McCormack, David C. Nagel, Sam L. Pool, M. Rhea Seddon, Joseph C. Sharp, and Frank M. Sulzman. January 1987. *Advanced Missions with Humans in Space.* No city of publication given: National Aeronautics and Space Administration.

Connors, Mary M., Albert A. Harrison, and Faren R. Akins. 1985. *Living Aloft: Human Requirements for Extended Spaceflight.* NASA SP-483. Washington, DC: National Aeronautics and Space Administration.

National Academy of Sciences. National Research Council. Aeronautics and Space Engineering Board. Commission on Engineering and Technology System. December 1987. *Space Technology to Meet Future Needs.* Washington, DC: National Academy Press.

National Academy of Sciences. National Research Council. Committee on Space Biology and Medicine. 1987. *A Strategy for Space Biology and Medical Science for the 1980s and 1990s.* Washington, DC: National Academy Press.

National Academy of Sciences. National Research Council. Committee on the Space Station. September 1987. *Report of the Committee on the Space Station of the National Research Council.* Washington, DC: National Academy Press.

National Aeronautics and Space Administration. August 1987. *Space Station Science Operations Management Concepts Study.* Washington, DC: National Aeronautics and Space Administration.

National Aeronautics and Space Administration. Office of Aeronautics and Space Technology. November 1987. Presentation on Project Pathfinder Technology Benefits Assessment.

National Aeronautics and Space Administration. Office of Space Science and Applications. Life Sciences Division. 1984. *Life Sciences Flight Program: Guide to the Life Sciences Flight Experiments Program.* No city of publication or publisher given.

National Aeronautics and Space Administration. Office of Space Science and Applications. Life Science Division. 1986. *Biological and Medical Experiments on the Space Shuttle 1981-1985.* Ed. Thora W. Halstead and Patricia A. Dufour. Washington, DC: National Aeronautics and Space Administration.

National Aeronautics and Space Administration. SESAC Task Force on Scientific Uses of Space Station. 1986. *Space Station Summer Study Report.* Ed. David C. Black and Hugh S. Hudson. No city of publication or publisher given.

National Aeronautics and Space Administration. Space Biomedical Research Institute. Lyndon B. Johnson Space Center. May 1987. *Results of the Life Sciences DSOs Conducted Aboard the Space Shuttle 1981-1986.* Ed. Michael U. Bungo, Tandi M. Bagian, Mark A. Bowman, and Barry W. Levitan. Houston: Lyndon B. Johnson Space Center.

National Aeronautics and Space Administration Advisory Council. Task Force on the Role of Man in Geosynchronous Orbit. February 1987. *Report of the Task Force on the Role of Man in Geosynchronous Orbit.* No city of publication given: National Aeronautics and Space Administration.

National Aeronautics and Space Administration Advisory Council. Space and Earth Science Advisory Committee. November 1986. *The Crisis in Space and Earth Science: A Time for a New Commitment.* No city of publication or publisher given.

National Aeronautics and Space Administration, American Astronautical Society, Baylor College of Medicine, National Institutes of Health, Universities Space Research Association, Sponsors. 1987. *Space Life Sciences Symposium: Three Decades of Life Science Research in Space.* Washington, DC, June 21-26.

National Commission on Space. May 1986. *Pioneering the Space Frontier.* New York: Bantam Books.

Physiologic Adaptation of Man in Space: VII International Man in Space Symposium, February 10-13, 1986, Houston, TX. Sponsored by National Aeronautics and Space Administration, Universities Space Research Association, Baylor College of Medicine, and International Academy of Astronautics. Ed. Albert W. Holland. *Aviation, Space, & Environmental Medicine* 58 (September 1987).

Pitts, John A. 1985. *The Human Factor: Biomedicine in the Manned Space Program to 1980.* NASA SP-4213. Washington, DC: National Aeronautics and Space Administration.

Ride, Sally K. August 1987. *Leadership and America's Future in Space: A Report to the Administrator.* Washington, DC: National Aeronautics and Space Administration.

U.S. Congress. House of Representatives. Committee on Science and Technology. October 1983. *National Aeronautics and Space Act of 1958, as Amended, and Related Legislation.* 98th Congress. 1st Session. Committee Print.

Thomas E. Malone, Ph.D.
Chairperson

Michael Collins

Francis D. Moore, M.D.

Beryl A. Radin, Ph.D.
Staff Associate

Program Administration

The coordination of all efforts related to life sciences at NASA is a complex activity. While prime responsibility for most of the program resides in the Life Sciences Division within the Office of Space Science and Applications (OSSA), other parts of NASA are indirectly involved in life sciences efforts. NASA offices related to this discipline include the Office of Aeronautics and Space Technology, the Office of Space Flight, the Office of Space Station, the Office of Commercial Programs, the Office of Management, and the Office of Equal Opportunity Programs. Program implementation requires coordination of efforts by different Headquarters and Center institutes. In most of the major program offices, the effort involves multiple divisions, and at the Ames Research Center (ARC) and Johnson Space Center (JSC), the main participating Centers, more than one Directorate.

Located along with the Life Sciences Division in OSSA are the Astrophysics, Solar System Exploration, Earth Science and Applications, Microgravity Science and Applications, and Communications Divisions. Organized into programmatic areas for Operational Medicine, Space Medicine and Biology, Flight Programs, and Biological Systems Research, the Life Sciences Division differs from its organizational peers in several ways: it emphasizes both manned and unmanned missions and projects and, perhaps most importantly, its conduct of science is closely linked with other parts of the NASA organization for certain programmatic efforts, as noted above.

This report emphasizes key findings for enhancing the effectiveness of life sciences at NASA. The analysis and the recommendations that follow are based largely on material collected during extensive interviews with multiple levels of program staff at NASA Headquarters, ARC and JSC, and other Federal agencies. Those interviewed included staff at the Office of the Administrator, OSSA, the Life Sciences Division Director's Office, branch and program officials in Washington, as well as Center staff in comparable positions at ARC and JSC. In addition, meetings were held with representatives from the National Institutes of Health (NIH), Office of the Secretary of Defense, and the Air Force, and correspondence was received from the National Science Foundation, the Department of Energy, and the Department of Agriculture. This paper reviews historical perspectives involving the Life Sciences Division, summarizes the current status of the program, and makes specific recommendations.

Issues

Historically, programs of the Life Sciences Division have been defined and constrained by a number of decision processes, several of which are discussed in this section. Some of these processes can be viewed as outcomes of the matrix organizational structure in which the programs are placed; others are defined by Agency- or Government-wide procedures; still others occur as a result of explicit choices by program officials.

The Budget Process

The NASA budget process is highly iterative and involves participants at all levels of the organization. Program managers work from the base of the previous year's budget, as well as the projection for the coming year contained within the past year's budget. Although this figure may shift as the process unfolds, the number serves as the perceived base line for the budget-planning effort. During the past few years, however, because of delays in the congressional budget process, NASA staff have been required to use budget estimates rather than appropriated funds as a planning base.

An overall view of the habitation modules for the Space Station is provided by this mockup at Marshall Space Flight Center.

Operating within the framework of the budget "mark" established by the Administrator's office, the Life Sciences Division engages in the development of the proposed budget along two decision processes mandated by the Agency: the Program Operating Plan (POP) and the Research and Technology Operating Plan (RTOP). The POP process has been used to establish priorities for flight projects within the Division for the coming year, focusing on specific projects and the time and resources required to implement those projects. To this point, the RTOP's have been devised through a process designed to establish research priorities in specific program areas; this process was used within the research area of the Division and frequently conceptualized as a multiyear research plan. Both of these processes have been developed through negotiations, visits, and reviews involving Center staff, program managers, and discipline scientists at NASA Headquarters. Although the balance between program elements does change somewhat from year to year, the Division's budget requests are split between funds for flight experiments and those for research and analysis.

A series of steps in the budget process proceeds to legislative appropriation: Centers, Division Director, OSSA, NASA Administrator, Office of Management and Budget (OMB), and Congress. At each step, the Life Sciences budget is subject to change, typically to decreases because proposed new projects are ultimately prioritized within many disciplines and not approved.

Over time, the Life Sciences Division has emerged with about 5 percent of the OSSA budget. The largest impacts on the proposed budget figure occurred before submission to OMB. During the past year, however, the Division's requests have been supported by NASA.

Additional but lesser cuts have usually been made in the proposed budget of the Life Sciences Division as it moves to OMB. At OMB, the decrease is generally part of a total reduction in the NASA budget.

Because congressional consideration of the budget is relatively well documented, especially through publication of hearings, it is possible to describe the treatment of the Life Sciences Division's proposal at the final level of decision. From FY 1983 through FY 1986, the NASA Administrator did not mention the Life Sciences program specifically in his statement submitted as an overview of the budget presentation. While other large projects were briefly described, no Division project was proposed.

Congressional actions on the Life Sciences programs reflected the budget requests made by the Agency for FY 1985. (The authorization process is separate from the appropriating process and represents another congressional perspective on the program.) The final authorization that emerged from the conference committee augmented various programs within OSSA. The budget for the Life Sciences Division was not, however, increased. These results were in contrast to the situations in 1977 and 1979, when budgets for Controlled Ecological Life Support Systems (CELSS) and Spacelab flight experiments were requested by NASA and approved by Congress.

Program Implementation

As the preceding discussion suggests, program implementation depends largely on factors influencing budget appropriations. As with many other Federal agencies, fiscal scarcity has been the driving force in much of NASA's program development, constraining the way that the organization has set forth plans for the future. Problems are compounded for programs of the Life Sciences Division because each participating office has its own budget, often won in competition with other offices involved in different projects and missions, and enjoys nearly complete discretion over how its funds will be spent.

Program planning for the Life Sciences Division is also complicated by uncertainties in flight opportunities. For most researchers involved with NASA, the real lure to participate in the Agency's work is the possibility of developing experiments on space flights. In the wake of the *Challenger* accident, as the

Agency reexamined its programs and procedures, these possibilities became more constrained and involved greater competition. In the succeeding months, changes in experiment manifests have become the rule, rather than the exception. Carefully crafted program plans, and national and international agreements developed to assure maximum yield from the already scarce flight resources, have been restudied, reprioritized, and reassigned.

Other challenges for the Life Sciences Division are caused by disconnections in many areas between ground-based research and space-flight experiment programs. When flight opportunities diminish, it is natural that more ground-based research is supported. As noted earlier, the POP process for Flight Programs is focused on specific projects amenable to traditional management rigor — timing, deadlines, and deliverables. By contrast, the science planning process (through RTOP's) uses a longer period of time; deadlines for results are usually inappropriate with this procedure. Most of the funding for ground-based research has been generated by unsolicited proposals from offerors acquainted with the programs in question. Space-flight experiments, on the other hand, have been selected through organized competitions and are not always ideally related to the ground-based program. Life Sciences Division staff are initiating the coupling of these activities through integrated project and program management.

In the past, the Division has found it difficult to plan and implement programs in an unstable environment. The uncertainty of a steady resource commitment has characterized the planning setting, making it challenging for the program to achieve an orderly implementation of strategies.

Headquarters and Center Roles

As indicated by the Phillips Committee, headed by former Apollo Program Director, General Samuel E. Phillips, post-*Challenger* NASA is not clear about the respective roles of the Centers and Headquarters. The Committee found that Headquarters program direction is not always firmly established — this is a special problem when the technical demands of some programs require contributions by more than one Center. The Committee's assessment of the technical requirements for long-duration flight emphasizes the need for more clearly defined roles to manage the competition among program components. The Life Sciences program shares the problem.

It is probably inevitable that there will be some level of competition between the Centers and Headquarters and among implementing Centers. From the perspective of the Centers, attempts by Headquarters to limit the autonomy of Center researchers and managers are a form of "micromanagement." They argue that Center researchers have had less autonomy than the academics who also receive NASA funds. From the perspective of Headquarters' staff, the Centers cannot operate as if they were separate NASA entities without policy direction and guidance from Headquarters. Because of competition among the Centers for major programs of the Life Sciences Division, Headquarters emphasizes the need to establish an overall framework to define the Centers' activities. A recent emphasis in the Division has been to develop complementary but integrated activities involving the Centers as well as Headquarters.

Personnel

The work force that carries out the NASA Life Sciences program is a composite of career civil servants, extramural scientists, and contractor staff. As with most research agencies, the program relies on external grantees and contractors to conduct NASA-funded research. Unlike most research agencies, however, the work that is conducted on NASA premises depends heavily on contractors and other external researchers. It is not unusual to find an activity with a few NASA career staff, many contractors, and university faculty or students who are using NASA facilities for their research.

As a result of various policies and practices, the permanent staff of the Life Sciences Division is a relatively homogeneous group, largely composed of individuals who have been in the organization for many years. This pattern is not unique to the Division; the average age of NASA personnel is 46 and increasing by almost a year per year. In a field where scientific changes occur rapidly, without the opportunity for new, younger hires, the aging staff is not always as current about recent developments as one might hope. Concern has also been expressed about the ability of the program to attract young scientists to its permanent staff when research opportunities and, hence, career development are subject to change.

Use of Outside Advice

Although the Life Sciences Division has called on advice from outside individuals and groups in a number of ways, two processes exist to structure the day-to-day use of outside advice — the peer-review process and the solicitation process.

Peer Review. The past procedures established within the Division to evaluate research proposals followed the separation between ground-based research and flight experiments. The ground-based research proposals, funded as "research and analysis," were evaluated by standing outside panels organized along the Division's program lines. Since 1965, these advisory panels have been formed and staffed by the American Institute of Biological Sciences (AIBS), an umbrella organization comprising more than 40 scientific societies and individual scientists.

The panels established for each of the major research and analysis program areas — Space Medicine, Space Biology, Exobiology, Biospherics, and CELSS — were constructed as standing multidisciplinary groups in which breadth and depth of knowledge are valued. Unlike peer panels in some other agencies, such as NIH, the groups reflect a heterogeneous rather than homogeneous slice of science. As a result, the panel's evaluation of proposals is based on two kinds of assessments: it depends on the one or two panel members who have specific expertise in the area proposed, and it calls upon the judgment of relative outsiders to the field to determine the relationship of the individual proposal to overall program goals. To this point, all research proposals — whether submitted by a NASA researcher or by an individual proposal in the outside research community — are reviewed by these panels.

In contrast to the procedures described above, the peer-review process for flight experiments identifies external reviewers from advisory committee and peer-review panels when the flight program solicits and receives proposals for experiments. As with the standing peer-review panels, AIBS manages the review. Both types of panels include intra- and extramural scientists.

Unique opportunities sometimes exist to develop flight experiments with abbreviated schedules and severe space and design constraints. In most cases, the experiments packaged for flight have undergone external peer review as earlier ground-based projects. Decisions to put a demonstration or test on a flight through a process called Detailed Supplementary Objectives are made at JSC. In those instances, the Center establishes a peer-review process that includes outside university participants, as well as a wide range of NASA scientists and managers.

Initiation of Proposals. The peer-review procedures described above are initiated when the Division receives proposals for research or flight experiments. Proposals can be initiated in four ways:

1) Request for Proposals (RFP) — a formal advertisement that requests specific services or products.

2) Announcement of Opportunity (AO) — a process used by NASA to request the submission of proposals addressing specific areas of research that NASA considers necessary to meet scientific objectives. All flight experiments of the Life Sciences Division must be solicited through the AO process.

3) Dear Colleague Letters — a semi-official procedure that announces opportunities for proposals in specific areas of focused research. The decision to send letters can be made within the Division as a means of providing information about NASA program goals and objectives. This could be replaced by a more formal process called a NASA Research Announcement.

4) Unsolicited proposals — a determination by NASA that it will respond to the priorities established by the academic community through its definition of appropriate research. Most of the Division's ground-based research and analysis programs, as well as most of the research and analysis programs within OSSA, have relied on unsolicited proposals.

External Relations

Historically, life sciences requirements have not been incorporated early enough into major NASA projects, with certain notable exceptions, such as the Viking Project and operational medicine activities. The accommodation of life sciences research requirements on the Space Station has been a difficult process. The lack of a specific call for life sciences specifications in the Space Station RFP suggested that the life sciences perspective had not been fully acknowledged. Programmatic interests have made it difficult to coordinate such activities as life support systems, human factors elements, and extravehicular activity (EVA). On the positive side, during the past few years the relationship between the Astronaut Corps and program of the Life Sciences Division has improved because of the mutual recognition of the importance of working together on an ongoing basis, rather than when it is time to put an experiment on flight.

Because of the nature of its programs, the Life Sciences Division has natural overlap with efforts under way in other Federal agencies. The Agency has made attempts over the years to develop contact with relevant research programs at the National Institutes of Health; these have had varying degrees of success. Recently, the White House Office of Science and Technology Policy has encouraged coordination among these programs.

Several other agencies and organizations (including the Department of Agriculture, the Department of Defense, specifically the Air Force, the Department of Energy, and the National Science Foundation) have expressed an interest in developing close, productive working relationships with NASA in their particular areas of interest and on efforts of mutual involvement. It was clear that those surveyed believe it essential that NASA programs be complementary or cooperative with other agencies sharing similar objectives. All representatives were amenable to meeting with NASA to explore available, appropriate mechanisms for furthering interagency collaboration. However, they emphasized that NASA, viewed as the leading Government agency in space biomedical research, should take the initiative in investigating such opportunities.

Findings and Recommendations

Most of the problems described above have been identified by staff of the Life Sciences Division, and major steps have been taken to address them. A significant recent effort was to assure that a properly constituted committee develop the long-range strategy and that implementation follows this step. The creation of the Life Sciences Division Science Management Plan in January 1988 was a significant milestone, for the document defines the major research component of the Life Sciences program and identifies the structural relationships among these elements. In this document, the Division has given attention to many issues that have confronted the program. At the same time, the magnitude of related challenges requires commitment and support by all levels of the NASA organization, not simply the Division.

Perspectives on NASA Life Sciences

Findings

- Through most of their existence, NASA life sciences programs have been viewed as level-of-effort activities within the Agency. Until very recently, they have experienced a number of problems that validate this general finding.

- The unique nature of these programs has been difficult for others both inside and outside the Agency to understand. Life sciences at NASA is both a centralized program (found in one organizational location) and a differentiated set of relationships spread throughout the Agency, affecting nearly every office in NASA. In the past, programmatic uniqueness was not well understood and, thus, the discipline suffered from low visibility and insufficient attention.

- The practice of dispersing various life sciences elements throughout the Agency has made it difficult for others to gain a sense of a visible and cohesive program.

- Life sciences activities have had difficulty gaining support in the budgetary process, being disproportionately affected when budget requests were dramatically reduced within NASA and often compounded by reductions by the Office of Management and Budget and Congress.

- Life sciences experiments rarely had high priority in the competition for access to missions. While life sciences issues were given considerable attention, they did not receive strong support from Agency personnel who determine access to missions. These issues were considered relevant to extended rather than short missions.

- NASA management tended to expect that most problems could be addressed through technical, engineering solutions and did not accept the fact that life sciences research has a long lead time to produce results.

- With few exceptions, life sciences does not have an organized and visible constituency to advocate its agenda with individuals who control resources.

- Joint efforts with other agencies as well as other parts of NASA are rare and receive little support from program administrators.

• In recent times, however, there are clear indications that many of these practices have changed or are in the process of changing, largely because of efforts by the Life Sciences Division to address these problems. Through specific activities by the Division that link life sciences efforts to the broader goals of the Agency, there is growing acknowledgement of the unique opportunities offered by the program.

- Growing support within NASA has been expressed through budget increases, backing within the Office of Space Science and Applications for capabilities such as an inflight variable-gravity facility, and increased visibility in planning activities, such as those undertaken by the Office of Exploration and the Agency-wide Management Planning Team. Similarly, intra-Agency cooperation with other NASA offices, such as those involving the human factors program within the Office of Aeronautics and Space Technology (OAST), signals a new visibility for life sciences within the Agency.

- The activity within the Life Sciences Division has also focused on cooperation with others outside the Agency. The program has renewed relationships with NIH, intensified its contact with international life sciences efforts, particularly those involving the Soviet Union, and stimulated new interest in the research community through efforts such as the Space Life Sciences Symposium, held in the summer of 1987.

— In each of these cases — especially those inside NASA — the Life Sciences Division has been the initiating player, calling on others to respond to its requests for participation in various decisions or decision-making arenas.

Recommendation

- Senior NASA administrators should clearly support Division efforts that link life sciences to the broader Agency goals by taking new actions, such as the following:

    — Formally acknowledging the important differences between life sciences and other science and engineering programs within NASA

    — Accentuating the importance of issues related to humans in space for the Agency's advanced missions

    — Institutionalizing the *ad hoc* efforts by the Division to be involved in Agency-wide planning (using existing processes, such as the program review, as a way of examining both the centralized and decentralized aspects of life sciences).

Life Sciences Goals

Findings

- Throughout most of its history, program goals within the Life Sciences Division have not been clearly articulated or disseminated. Until very recently, the Life Sciences program has been an effort that was simply the sum of individual parts, with the disparate pieces standing or falling on their own. Moreover, since the lunar landing, there had not been a vision of the future uniting the individual program pieces and providing a convincing justification of the expenditure of time and money.

- During the past 2 years, and specifically during the period of the Life Sciences Strategic Planning Study Committee (LSSPSC) effort, the Life Sciences Division has made great strides in addressing these past practices. Responding to the general recommendations given in *A Strategy for Space Biology and Medical Science for the 1980s and 1990s* (National Academy of Sciences, 1987), as well as to suggestions by the LSSPSC, the establishment of the system for developing Program Disciplinary Plans holds great promise. These plans will include both ground and flight research activities, as well as intramural and extramural research. Once developed and disseminated, the process will provide a vehicle for others to comprehend the Division's program goals.

Recommendations

- Senior NASA management should support current efforts to enunciate Life Sciences program goals and should provide stable policy and fiscal support for the Life Sciences Division that will allow these initial planning efforts to develop and continue.

- The Life Sciences Division should assist the disciplinary groups and senior management by anchoring these efforts within a broader framework — developing overall goals for the Division that reflect the alternative long-range plans now being considered for the Agency.

Life Sciences Organization

Findings

- Throughout its history, NASA's total effort related to life sciences has been complex and extremely fragmented in terms of organization structure and decision processes.

    - This fragmentation has resulted from the complexity of the program effort and NASA itself. Implementation of life sciences research involves some degree of effort by every major NASA program office, particularly those responsible for the NASA Centers.

    - This fragmentation of responsibility makes it more difficult to coordinate the activities of scientists and administrators in the Centers and those in NASA Headquarters.

    - In addition, life sciences efforts must balance the separate imperatives of flight and ground research, as well as the differing perspectives of inside and outside researchers.

- At the present time, the Life Sciences Division has worked to manage the differing but complementary perspectives of these various participants.

    - The newly developed Program Disciplinary Planning process seeks to integrate the ground and flight, intramural and extramural, and international components of the program. Through this process, the roles of the Centers and Center staffs will be clarified and meshed with research plans involving outside scientists.

    - If funds are available, the program plans to institute Specialized Center of Research (SCOR) efforts.

    - The reorganization of the Flight Programs office also clarifies the appropriate levels of overlap between flight and ground activities and, at the same time, emphasizes the special program and project nature of flight efforts.

- The creation of the Life Sciences Senior Management Council, including senior Center staff as well as key managers from Headquarters, provides a forum for discussion and a mechanism to resolve program-wide issues.

Recommendation

- **Senior NASA management should play a more active role in supporting efforts of the Life Sciences Division to institutionalize substantive linkages with relevant program elements in other parts of the Agency.**

NASA Life Sciences Advisory System

Findings

- Although the Life Sciences program has historically relied on individuals and groups outside the Agency for advice and support, it has had difficulties establishing stable partnerships.

  - Through much of its life, the program has called on outside scientists and consultants in a variable and often unpredictable way. It was not always clear to program managers how they could adapt outside advice concerning life sciences requirements and NASA realities. Too frequently, the recommendations of blue ribbon scientific advisory committees could not be accomplished because NASA staff could not find a way to implement them within organizational, budget, and personnel limitations.

  - Similarly, program managers were not clear about how to develop a constituency of support for the new ideas that filtered into the Agency or how to build an inchoate constituency into a coordinated and productive program. Budget limitations encouraged program managers to avoid outreach to new constituents who, while potential supporters, were also supplicants for very scarce research dollars.

- The development of a new advisory structure for the Life Sciences program supported by Discipline Working Groups holds great promise as a way of addressing many of these historical problems.

  - The creation of Discipline Working Groups, including both outside scientists as well as Center scientists, provides a framework for the effective use of cohesive scientific advice in program development activities.

  - These groups will be part of the program disciplinary planning process, with a separate group of outside scientists serving as the mechanism for peer review of proposals. The chairs of the groups will constitute a Division Science Working Group.

Recommendation

- **The Life Sciences Division should evaluate the new advisory process, which represents a significant and positive step, as soon as the new process is instituted.**

Working Relationships Between NASA Life Sciences and Other Groups Inside and Outside the Agency

Findings

- The Life Sciences Division has had variable success working with other domestic and international organizations involved in space life sciences research. Too frequently, the program operated as an isolated effort, avoiding relationships with other research groups. This isolation was expressed in terms of its relationships with other scientists inside NASA, in other agencies, in the broader university community, and in international activities.

  - In the past, the Division had difficulty forging working relationships with other offices at NASA Headquarters. The debate about an inflight variable-gravity facility for the Space Station illustrated this problem. Lack of coordination on such issues as life support systems and human factors elements also was evidence of this difficulty.

  - Because of the nature of its programs, the Life Sciences Division has natural overlap with efforts under way in other Federal agencies. As noted earlier, NASA has made efforts over the years to develop contact with relevant research programs at the National Institutes of Health; recently, the White House Office of Science and Technology Policy expressed concern about the lack of coordination among these programs. Activities within the Department of Agriculture, the Department of Defense, particularly the Air Force, the Department of Energy, and the National Science Foundation also are relevant to Division efforts.

  - In addition, the Life Sciences Division has had variable success in forging relationships with universities and other research institutions essential to the development of an ongoing research community. While some training programs were in operation, they tended to be very small *ad hoc* efforts that did not provide a mechanism to bring young investigators into the system. Some NASA scientists were involved with neighboring universities and research institutes, but these efforts were not systematically encouraged.

  - Past cooperation between programs of the Life Sciences Division and related international efforts has been more positive. However, at times these efforts were not closely linked to other parts of the Division's activities.

- The Division has adopted a strategy that attempts to increase the visibility of its programs and collaborative arrangements with other scientific groups.

  - Within NASA, an agreement has been reached to work with OAST on space human factors efforts. Similar agreements have been reached with the Office of Space Station for the Health Maintenance Facility and environmental requirements for the Space Station and with the Office of Space Flight to manage missions from a medical perspective.

- A joint funding arrangement is in the process of development with NIH for SCOR grants, and conversations have been held with other Federal agencies.

- NASA Center staff have been encouraged to develop relationships with universities in their geographic areas.

- International collaboration — particularly activities involving the Soviet space program — has been intensified.

Recommendations

- **The Life Sciences Division should increase its outreach activities to the broader scientific community and develop strategies and implementation plans that grow out of the program recommendations included in this report, as well as the specific plans that emerge for the program disciplinary planning process.**

- **NASA should develop both policy and financial support for new relationships with universities, encouraging joint appointments at NASA Centers and local universities in specific research areas and providing funds and new mechanisms for the training of young scientists. In addition, the Agency should establish professorships in space life sciences at selected universities.**

- **Senior personnel from the Life Sciences Division should participate in all top-level planning of Agency programs.**

- **International collaboration should also be increased by providing reciprocal training opportunities for individuals at the Centers.**

Staffing for NASA Life Sciences

Findings

- The permanent staff of the Life Sciences Division is a relatively homogeneous group, largely composed of individuals who have been in the organization for many years.

  - As a result of constraints imposed by budget limitations as well as policy determinations, the program could not hire new, younger personnel. In addition, the unpredictable nature of opportunities for flight research has made it difficult to attract young scientists to the permanent staff. This is especially problematic in a field where scientific changes occur rapidly.

  - Moreover, important leadership positions in the program have remained unfilled for long periods of time.

- Budget and personnel constraints have forced the program to depend heavily on contractors to supplement the civil service staff. Constrained by available funds, the program rarely used the short-term possibilities for appointment available through the Intergovernmental Personnel Act (IPA) and the potential for loan of scientists from other Federal agencies.

- During the past year, much has changed.
  - New slots have been created at the Centers and Headquarters, and some of the vacant positions have been filled.
  - The Centers have been encouraged to use the opportunities available through the IPA mechanism, as well as the loan of scientists from other Federal agencies.
  - Increased attention has been given to ways of expanding training programs.

Recommendation

- **The Life Sciences Division should continue to address staffing problems and call on senior NASA management to support this effort. In addition, the Division should have a formal mechanism for both long- and short-term training to develop a new generation of top-quality scientists on the permanent NASA staff.**

Peter B. Dews, M.D.
Chairperson

Carolyn L. Huntoon, Ph.D.

Frederick C. Robbins, M.D.

Mark Schlam
Staff Associate

Mathew R. Schwaller, Ph.D.
Staff Associate

Applications

NASA's space program is built on a history of innovation, research, and development in science and engineering. Although applications research is not a part of the Agency's primary goal in space exploration, many of NASA's innovations do have commercial potential. Indeed, NASA's programs have generated over 30,000 documented spinoffs.

It is clearly in the national interest to transfer NASA's technological innovations to the private sector. Technology transfer and commercialization can, however, divert resources from the overall Agency mission. Thus, NASA has had to strike a balance between its primary mission in space exploration and its interest in applying research results to new products and services. This summary examines the policies that govern NASA's ability to develop applications and to transfer relevant technology. In addition, it explores current Life Sciences programs to identify areas where technological innovations are likely to yield significant new commercial applications in the near term.

Federal Policy Concerning Space Applications

The Space Act of 1958 established NASA as the agency responsible for the U.S. space program and recognized the importance of space exploration in areas of national interest, such as defense, economic development, and scientific competitiveness. It also required NASA to promote the peaceful use of space for the benefit of mankind. The Stevenson-Wylder Technology Innovation Act of 1980 went further in defining how these objectives can be met, specifically, by funding programs to transfer innovative space technology into the non-space sectors of U.S. society. This legislation created the Federal Laboratory Consortium to encourage the exchange of scientific and technical personnel among Government-funded laboratories and to establish Commercial Centers for the Development of Space. These centers, identified in table 2, now serve as focal points for innovative research and development related to space by providing seed money and technical advice to promising commercial ventures, especially to small businesses.

The Federal Technology Transfer Act of 1986 made technology transfer a responsibility of each scientist and engineer at Federal laboratories and a factor to be considered in promotion policies, performance evaluations, and job descriptions.

Among other matters, the act mandated a minimum 15-percent royalty to be paid to inventors for their licensed innovations, established a cash awards program to reward scientists and engineers for their innovations, and established a Federal Laboratory Consortium to assist in advising, training, and promoting technology transfer. President Ronald Reagan summarized the accomplishments of the legislation as follows:

> A vigorous and technological enterprise involving universities, industry and government laboratories is essential to our economic growth and national security . . . . With the Federal Technology Transfer Act of 1986 . . . the government has removed many of the barriers to industrial use of publicly funded technological research.

Table 2. *The 16 Commercial Centers for the Development of Space, Their Host Facilities, and the Year of Their Inauguration*

1. Center for Advanced Materials, Battelle Columbus Laboratories, Columbus, Ohio. 1985.
2. Center for Advanced Space Propulsion, University of Tennessee Space Institute, Tullahoma. 1987.
3. Center for Bioserve Space Technologies, University of Colorado, Boulder. 1987.
4. Center for Cell Research, Pennsylvania State University, University Park. 1987.
5. Center for the Commercial Development of Autonomous and Man-Controlled Robotic Sensing Systems in Space, Environmental Research Institute of Michigan, Ann Arbor. 1987.
6. Center for the Commercial Development of Space Power, Auburn University, Auburn, Alabama. 1987.
7. Center for Commercial Development of Space Power, Texas A&M Research Foundation, College Station, Texas. 1987.
8. Center for Development of Commercial Crystal Growth in Space, Center for Advanced Materials Processing, Clarkson University, Potsdam, New York. 1986.
9. Center for Macromolecular Crystallography, University of Alabama - Birmingham. 1985.
10. Center for Mapping, Ohio State University, Columbus. 1986.
11. Center on Materials for Space Structures, Case Western Reserve University, Cleveland, Ohio. 1987.
12. Center for Space Automation and Robotics, University of Wisconsin - Madison. 1986.
13. Center for Space Processing of Engineering Materials, Vanderbilt University, Nashville, Tennessee. 1985.
14. Center for Space Vacuum Epitaxy, University of Houston, Texas. 1986.
15. Consortium for Materials Development in Space, University of Alabama - Huntsville. 1985.
16. ITD Space Remote Sensing Center, NASA National Space Technology Laboratories, Mississippi. 1985.

NASA Policy on Applications and Technology Transfer

Since its inception, NASA has pioneered in technology transfer and applications research and has led Government agencies in this effort. A separate NASA Office of Applications was first established in 1971. Following a 1984 reorganization, the Office of Commercial Programs (OCP) was formed to disseminate technical information and to encourage technology transfer into the commercial sector. To meet these goals, the OCP sponsors seminars and meetings to acquaint non-NASA personnel with potential applications of NASA technology to industry, it manages the Technology Utilization Program, which maintains offices and technology applications teams at each field center, and it funds the Scientific and Technical Information Facility. These organizations serve as active transfer agents of NASA technical information, and they promote applied engineering. Another OCP responsibility is facilitating the flow of technical information from NASA laboratories through a series of publications, including "Tech Briefs," which describe NASA-developed innovations in concepts, devices, and processes; *Spinoff*, which reports on a selection of products derived from NASA technology; and the *Patent Abstracts Bibliography*, which lists NASA inventions. In addition, the OCP provides computerized access to various NASA data bases and computerized networks to link various technology utilization centers. The OCP also provides financial support to the Commercial Centers for the Development of Space, identified above. The office recently established three university centers in the life sciences, as discussed in the next section.

Applications Research and Technology Transfer in Life Sciences

The Life Sciences Division at NASA Headquarters is the organizational entity primarily responsible for life sciences program planning and development. Currently, there are no Division staff specifically assigned to review projects for applications possibilities, nor does anyone in the Division represent Life Sciences in the process of technology transfer. Support for life sciences applications research and technology transfer relies primarily on the NASA field centers, on OCP efforts, as well as on projects supported by the Space Station Office. Certain Life Sciences programs do, however, have commercial potential.

Commercial Centers Established by the Office of Commercial Programs

The OCP provides funding of up to $1 million to establish commercial development centers. NASA also offers scientific and technical expertise to these centers, as well as opportunities for cooperative activities and other forms of continuing assistance. Additional funding comes from corporate and university affiliates, which are expected to increase their support to sustain the centers after a period of 5 years.

The paragraphs below identify the three university centers established in the past 3 years that are specifically concerned with life sciences applications and technology transfer.

Center for Macromolecular Crystallography, University of Alabama, Birmingham. This center, established in 1985, specializes in microgravity crystal growth of biological materials identified by participating firms in the pharmaceutical, biotechnology, and chemical industries. The center's goal is perfection of the technology for space-based material processing of biological crystals.

Center for Cell Research, Pennsylvania State University, University Park.
The major goals of this center, created in 1987, are as follows: 1) designing and testing methods for manipulating cell secretions on Earth and in space, 2) increasing the production of selected secretory molecules, and 3) reducing the cost of producing and purifying commercially valuable cell secretions by using space-based techniques.

Center for Bioserve Space Technologies, University of Colorado, Boulder. This center, instituted in 1987, has four main objectives: 1) pharmaceutical testing in microgravity; 2) production and evaluation of various bioproducts, such as natural and synthetic skin, cartilage, and lenses for the human eye; 3) production and evaluation of specialized, biologically active membranes; and 4) testing new high-yield agricultural strains in space.

Projects Supported by the Space Station Office

The Office of Space Station has established an active program to encourage commercial applications and to facilitate technology transfer in several areas, including the life sciences. Commercial applications are being pursued by the Space Station Utilization Office through two working groups on life sciences applications and technology transfer: the Space Station Commercial Advocacy Group and the Life Sciences Commercial Working Group. These groups have identified the mission requirements for life sciences experiments of commercial interest, including electrophoretic separation of biological compounds and protein crystal growth. These requirements will be considered in the final designs of the Space Station infrastructure. In addition to providing a healthy environment for commercial development in space, the Office of Space Station also supports innovative programs that may yield significant new applications in the near term. One such program is the Health Maintenance Facility (HMF) for the Space Station.

The HMF is designed as a multipurpose inflight clinic on the Space Station that will serve four goals:

- Ensure crew safety and health maintenance during routine operations
- Prevent early mission termination due to medical conditions
- Prevent unnecessary rescue
- Ensure the probability of success of a necessary rescue.

No mission in space can be risk free, but the goal for the HMF is to anticipate health risks and to provide countermeasures that can reduce risk to a well-defined and an acceptable level. It accordingly has capabilities for prevention, in part with exercise facilities, for diagnosis, and for treatment, including care for acute health problems. Some of these capabilities, as they are refined through experience, may well have commercial applications, as noted in the next section.

Applications Potential of Life Sciences Programs

The Life Sciences Division does not specifically support any applications research or technology transfer projects, but many of the efforts in Operational Medicine, Space Medicine and Biology, Flight Programs, and Biological Systems Research have near-term or long-term commercial potential.

Biomedical Research Program. In conjunction with the Health Maintenance Facility sponsored by the Space Station Office, a research program was developed by the Life Sciences Division to define the operational requirements of the HMF and to conduct the research necessary to support its design and implementation. Developments are anticipated in analytical and surgical techniques and in noninvasive diagnostic measures, such as digital imaging of hard and soft body tissue. The HMF also offers new opportunities to adapt anesthetics, sterile manipulation devices, and drugs to combat the deleterious effects of space flight. A number of these and other practices developed by the Biomedical Research Program may find application in prevention, diagnosis, and treatment of disease on Earth.

Other Scientific Programs. The Life Sciences Division supports a number of scientific programs, including those in exobiology, global biology, and the physiology of plants and animals. The investigations under way in these disciplines are primarily basic research. Generally, no near-term commercial applications are obvious in these areas, although applications with enormous usefulness to society may emerge at any time, often in quite unanticipated ways, as can happen with all basic research. The understanding built on scientific discovery serves as the foundation for technical innovation, technology transfer, and commercially viable applications.

Findings and Recommendations

Finding

- Although NASA's primary goal is space exploration, the Agency has a long history in applications research and technology transfer. Public and commercial sectors of the Nation have given considerable support to programs that adapt innovations from the space program to the national defense and to new commercial products and services. The primary responsibilities of the NASA Life Sciences programs are in scientific and biomedical research, not in applications research and technology transfer. While the

Agricultural practices show up as patterns of circles and squares in this Landsat 5 image of Garden City, Kansas. Squares are fields of crops, while circles result from the practice of center pivot irrigation. Color variations can be attributed to differences in crops and to different stages of crop maturity.

Division does not have a specific applications program, life sciences personnel at the Centers have communicated effectively with private enterprise.

Recommendation

- **The Life Sciences Division should continue to cooperate with private enterprise to help build awareness, interest, and support for the Division's research and development efforts. This should be accomplished primarily through Center personnel. The Division should consider identifying a staff member at NASA Headquarters as the focal point for the receipt and referral of suggestions for applications.**

Finding

- The Life Sciences Division supports programs, such as the Biomedical Research Program, that may generate commercially viable applications in the near term. Life sciences applications research and technology transfer at NASA are principally supported by the Office of Commercial Programs through publications and special projects and through the Commercial Centers for the Development of Space. Other significant applications research and technology transfer programs in life sciences include the Health Maintenance Facility and the working groups on commercial applications, both of which are supported by the Space Station Office.

Recommendation

- **The Life Sciences Division should continue to cooperate closely with other NASA offices that support applications research and technology.**
    - **In addition, Division representatives should have an advisory role in the Commercial Centers for the Development of Space in the Life Sciences.**
    - **The Division should take an active role in the Life Sciences Commercial Group and in supporting the Scientific and Technical Information Facility.**

Appendix

Appendix

Background on the Committee

The activities of the Life Sciences Strategic Planning Study Committee (LSSPSC) cover a relatively brief period. Established during spring 1986, the Committee convened initially in September 1986. It concluded its work 20 months later, in March 1988. The product of the LSSPSC's efforts is this report.

To meet its tasks, outlined in the Foreword, the Committee organized itself into Study Groups, each consisting of two to five Committee members and one to three consultants identified as Staff Associates. Of the original 11 Study Groups, 6 corresponded to NASA programs: Biomedical Research, Operational Medicine, Gravitational Biology, Controlled Ecological Life Support Systems (CELSS), Biospherics, and Exobiology. The remaining five Study Groups investigated issues that reached across program lines and scientific disciplines: Radiation, Systems Engineering, Infrastructure, External Relations, and Applications. Each group was charged with studying its given topic and recording its findings, along with corresponding recommendations, in a white paper. The findings and recommendations of the white papers were used as the basis for the overall findings and recommendations advanced by the Committee.

The original organization and tasking worked effectively, requiring only a few modifications. When it became apparent that the scope of Systems Engineering was too broad for one Study Group, two additional groups were added: Crew Factors and Flight Programs. Figure 4 lists the resulting 13 Study Groups. Along with the chairpersons and Committee members, it identifies the Staff Associates, who participated with the Study Group members in researching and drafting the white papers.

As plans developed for the final report, the Committee decided to incorporate the findings and recommendations of the External Relations Study Group into the "Program Administration" paper. This material included information elicited from a letter sent to 480 principal investigators in the larger scientific community informing them of the Committee's study and requesting suggestions concerning research and development, as well as administrative procedures, in the space life sciences. Figure 5 provides a copy of this letter, which drew about 100 responses. A summary of the comments and a list of the respondees appears at the end of the presentation on "Background on the Committee."

The LSSPSC met six times to review progress toward its final report. Figure 6 identifies these meetings by date, place, and agenda highlights. The first three

Appendix

meetings were informational and designed to orient Committee members to key life sciences issues. The last three involved reviews of Study Group papers and then successive iterations of the Committee's report. At its last session, on March 11, 1988, the LSSPSC formally approved the draft report and concluded its activities.

Study Group Assignments

Study Group	Chairperson	Study Group Members	Staff Associate(s)
Scientific and Technical			
Biomedical Research	Bernadine Healy, M.D.	William DeCampli, M.D., Ph.D. Frederick C. Robbins, M.D.	Warren Lockette, M.D.
Radiation	William DeCampli, M.D., Ph.D.	Francis D. Moore, Ph.D.	Mark H. Phillips, Ph.D.
Crew Factors	William C. Schneider, D.Sci.	Gerald P. Carr, P.E., D.Sci. Michael Collins	Lauren Leveton, Ph.D.
Systems Engineering	William C. Schneider, D.Sci.	Gerald P. Carr, P.E., D.Sci. Michael Collins Peter B. Dews, M.D. Jay P. Sanford, M.D.	Lauren Leveton, Ph.D.
Operational Medicine	Jay P. Sanford, M.D.	Ivan L. Bennett, M.D. Carolyn L. Huntoon, Ph.D.	Barry J. Linder, M.D.
Biospherics Research	Peter M. Vitousek, Ph.D.	Sherwood Chang, Ph.D.	Mathew R. Schwaller, Ph.D
Exobiology	Sherwood Chang, Ph.D.	J. William Schopf, Ph.D.	Mitchell K. Hobish, Ph.D.
Gravitational Biology	J. William Schopf, Ph.D.	Arthur W. Galston, Ph.D.	Keith Cowing
CELSS	Arthur Galston, Ph.D.	Peter M. Vitousek, Ph.D	Ross Hinkle, Ph.D.
Flight Programs	William C. Schneider, D.Sci.	Gerald P. Carr, P.E., D.Sci. Michael Collins	Keith Cowing Mitchell K. Hobish, Ph.D. Lauren Leveton, Ph.D.
Applications	Peter B. Dews, M.D.	Carolyn L. Huntoon, Ph.D. Frederick C. Robbins, M.D.	Mark Schlam Mathew R. Schwaller, Ph.D.
Institutional			
Infrastructure	Thomas E. Malone, Ph.D.	Michael Collins Francis D. Moore, M.D.	Beryl Radin, Ph.D.
External Relations	Ivan L. Bennett, M.D.	Bernadine Healy, M.D. Robert H. Moser, M.D.	Carole O'Toole

Figure 4. *The LSSPSC organized into 13 Study Groups to conduct its work.*

CASE WESTERN RESERVE UNIVERSITY · CLEVELAND, OHIO 44106

Dear

As you may know, NASA is now developing wide-ranging plans covering its space activities for the rest of the 20th century and the early part of the 21st. The programs that are under consideration present dramatic new possibilities for research in space and may require extensive new knowledge about human capacity to adapt to the space flight environment.

To assist in planning these future programs, NASA has organized a Life Sciences Strategic Planning Study Committee (LSSPSC). The Committee's mission is to recommend major goals for the Agency's activities in the life sciences and to lay out approaches for the attainment of these goals. The Committee is of the opinion that it is very important in order to prepare a more meaningful report that the points of view of those conducting space related research and the various relevant organizations be solicited.

As chairman of the LSSPSC, I am writing to make you aware of the Committee's undertaking and to solicit your views on the space program's past and future involvement with the life sciences. Specifically, any suggestions for both ongoing and proposed research and development work in the life sciences that you believe should be supported by NASA during the coming decade will be welcome and will receive careful consideration in formulating the Committee's recommendations. Also, any opinions on how NASA might improve communications with the scientific community concerning its programs and objectives in the medical, biological, and behavioral sciences, and any suggestions you might have for improving contracting/granting procedures would be useful to the Committee in meeting its charge.

Cleveland Study of the Elderly
Department of Epidemiology
  and Biostatistics
School of Medicine
Area Code: 216 Telephone 368-3760

Figure 5. *The External Relations Study Group circulated a form letter to researchers and administrators in the life sciences to collect information for the LSSPSC report.*

Appendix

```
        If possible, please respond by April 30, 1987, to:

                Dr. James Bredt
                Executive Secretary
                Life Sciences Planning Study Committee
                National Aeronautics and Space Administration
                Room 300 (Mail Stop EBR)
                600 Independence Avenue, SW
                Washington, DC  20546

Your comments will remain confidential should you desire.

On behalf of the Committee, I thank you for taking time to furnish us with
your comments.

Sincerely,


Frederick C. Robbins
Chairman,
  NASA Life Sciences Strategic
  Planning Study Committee
```

Figure 5. *The External Relations Study Group circulated a form letter to researchers and administrators in the life sciences to collect information for the LSSPSC report (continued).*

LSSPSC Meetings

Meeting 1

- Date and Place
  - September 24-25, 1986
  - NASA Headquarters, Washington, DC

- Agenda Highlights
  - Overview of current life sciences activities within NASA and past NASA advisory committee activities
  - Discussion of charge to LSSPSC
  - Organization into Study Groups

Meeting 2

- Date and Place
  - January 22-23, 1987
  - Johnson Space Center (JSC), Houston, TX

- Agenda Highlights
  - Tour of JSC facilities for life sciences research
  - Presentations on JSC life sciences activities
  - Progress reports by Study Groups

Meeting 3

- Date and Place
  - April 29-30, 1987
  - Ames Research Center (ARC), Moffett Field, CA

- Agenda Highlights
  - Tour of ARC facilities for life sciences research
  - Presentations on ARC life sciences activities
  - Study Group presentations of white paper outlines

Figure 6. *The Committee convened six times to review its progress.*

Meeting 4

- Date and Place
  - August 17-18, 1987
  - Boston Park Plaza Hotel, Boston, MA
- Agenda Highlights
  - Study Group reports on draft white papers
  - Discussion of possible provisional recommendations to NASA
  - Discussion of plans for completion of Study Group papers and LSSPSC report

Meeting 5

- Date and Place
  - November 20-21, 1987
  - Science Applications International Corporation (SAIC), McLean, VA
- Agenda Highlights
  - Discussion of draft 1 of LSSPSC report
  - General review of Study Group white papers

Meeting 6

- Date and Place
  - March 11, 1988
  - Jet Propulsion Laboratory, Pasadena, CA
- Agenda Highlights
  - Discussion and acceptance of final draft of LSSPSC report
  - Conclusion of Committee activities

Figure 6. *The Committee convened six times to review its progress (continued).*

Summary of Responses to Letter Circulated by External Relations Study Group to Private Industry and Academia

Abstract

Three basic emphases emerged in the approximately 100 responses to the Committee Chairman's letter of April 17, 1987. The respondents generally endorsed special interest research goals, suggested changes to enhance funding procedures, and recommended increased access to research facilities in space.

Research Topics

Specialists from a large number of disciplines responded to the letter. Many recommended continued or expanded research activities in areas of special interest to the Life Sciences. Discipline areas identified by respondents included the following:

> radiobiology, clinical diagnosis and treatment, cell and tissue culture, plant biology and physiology, evolution of life, digestive physiology and nutrition, bone demineralization and recovery, Controlled Ecological Life Support Systems (CELSS), Mars mission, Search for Extraterrestrial Intelligence (SETI), cell and molecular biology, cardiovascular and musculoskeletal systems, psychology and sociology of weightlessness and isolation, anti-emetic drug therapy, dental restoration in microgravity, space radiation carcinogenesis, acupuncture therapy for space motion sickness, immune system effects of microgravity, calcium metabolism, and exobiology.

Funding

Virtually no respondents emphasized the need for large and immediate increases in Life Sciences funding for specific projects. This was probably attributable to somewhat lowered funding expectations in view of Federal budgetary restraints. Three specific concerns were raised, however. (1) Many respondents felt the need for changes in the procedure for research proposal review, and several suggested panel review to help reduce the inbreeding often associated with peer review. (2) Many respondents suggested increased advertising of Announcements of Opportunity (AO's) and Requests for Proposals (RFP's) in national scientific journals. (3) One respondent made a strong case that NASA needs a policy of firm financial commitment to multiyear programs. It is difficult to plan multiyear studies if program managers reduce second and third year awards by more than 5 percent.

Flight Missions

Respondents with broad views of the Life Sciences program addressed their comments to the crucial problem of access to space. One respondent made the point that "scientific excellence demands rigorous results and high productivity." He, along with others, thought that the greatest impediment to productivity in the Life Sciences continues to be the limited access to microgravity.

Respondees to Letter Circulated by External Relations Study Group

E. John Ainsworth
Lawrence Berkeley Laboratory
University of California

Richard R. Almon, Ph.D.
Department of Biological Sciences
State University of New York at Buffalo

Mr. Sean Amour
Neurocybernetics Research Institute

Col. George K. Anderson, President
Society of U.S. Air Force Flight Surgeons

Robert S. Bandurski, Ph.D.
Department of Botany and Plant Pathology
Michigan State University

Mary Anne Bassett Frey, Ph.D.
The Bionetics Corporation

M.A. Benjaminson, Ph.D.
New York College of Osteopathic Medicine

Daniel D. Bikle, M.D., Ph.D.
Veterans Administration Medical Center

Frank D. Booth, Ph.D.
Department of Physiology
University of Texas Medical School

Allan H. Brown, Ph.D.
Department of Biology
University of Pennsylvania

Charles E. Bugg
American Crystallographic Association

John Carey, President
American Oceanic Organization

Arland L. Carsten, Ph.D.
Associated Universities, Inc.
Brookhaven National Laboratory

Morris G. Cline
Department of Botany
Ohio State University

Augusto Cogoli, Ph.D.
Swiss Federal Institute of Technology

George Crampton, Ph.D.
Department of Psychology
Wright State University

Diane Damos, Ph.D.
Institute of Safety and Systems Management
University of Southern California

Hector F. DeLuca, Ph.D.
Department of Biochemistry
University of Wisconsin

Richard M. Dillaman
Institute for Marine Biomedical Research
University of North Carolina

Paul A. Ebert
American College of Surgeons

Frederick R. Eirich
Polytechnic Institute of New York

Joseph J. Eller, M.D., Director
Pan American Medical Association

Ray Evert, Ph.D.
Department of Botany
University of Wisconsin at Madison

Ken Fisher
LSRO/FASEB

J. Charles Forman, Executive Director
American Institute of Chemical Engineers

Sidney W. Fox
Institute for Molecular and Cellular Evolution
University of Miami

E. Imre Friedmann
Department of Biological Sciences
Florida State University

Dr. Louis Friedman, Executive Director
Planetary Society

C.A. Fuller
Department of Animal Physiology
University of California

G. Robert Gadberry, Executive Vice President
American Cancer Society

Joel P. Gallagher
Department of Pharmacology and Toxicology
University of Texas

Harry K. Genant, M.D.
Department of Radiobiology
University of California

Roy Gibson, Ph.D.
British National Space Centre

Thomas J. Ginley, Ph.D., Executive Director
American Dental Association

Jay M. Goldberg, Ph.D.
Department of Pharmacology
University of Chicago

Victor M. Goldberg, M.D.
Orthopedic Research Society

A.W. Goode, M.D.
The London Hospital (White Chapel)

Dr. S. Graham
School of Medicine
State University of New York at Buffalo

Ralph R. Grams, M.D.
Department of Pathology
University of Florida

Dr. B. Gregor, Secretary
Geochemical Society
Department of Geological Sciences
Wright State University

Rufus K. Guthrie
School of Public Health
University of Texas

C. Rollins Hanlon, M.D., Director
American College of Surgeons

W. Darryl Hansen, Executive Director
Entomological Society of America

Hyman Hartman, Ph.D.
Massachusetts Institute of Technology

J.M. Hayes, Ph.D.
Biogeochemical Laboratories

Background on the Committee

Norman K. Hollenberg, M.D.
Department of Radiology
Brigham and Women's Hospital

Bruce A. Houtchens, M.D.
Department of Surgery
University of Texas

William Irvine
Radio Astronomy
University of Massachusetts

Gilbert Janauer
Department of Chemistry
State University of New York
  at Binghamton

Webster S.S. Jee, Ph.D.
Division of Radiobiology
University of Utah

Arthur Johnson
Alliance for Engineering in
  Medicine and Biology

Thomas H. Jukes
Space Science Laboratory
University of California

Gertrude Jungmann, Secretary-Treasurer
Institute for Gravitational Strain
  Pathology

Dr. R. Kahn
American Diabetes Association, Inc.

Nick Kanas, M.D.
Veterans Administration Medical Center

Peter B. Kaufman, Ph.D.
Division of Biological Science
The University of Michigan

Dr. F. J. Kloche
American College of Cardiology

K.L. Koch
Hershey Medical Center
Pennsylvania State University

Peter J. Lang, Ph.D.
Department of Clinical Psychology
University of Florida at Gainesville

Dr. L. Lemberger
ASPCP

C. Lenfant, M.D.
Public Health Service
National Institutes of Health

George Malacinski, Ph.D.
Department of Biology
University of Indiana

Dr. G.M. Martin, President
Department of Pathology
University of Washington

Gordon A. McFeters
Department of Microbiology
Montana State University

Dr. Jay Moskowitz
Associate Director of Scientific
  Program Operations
National Heart, Lung, and Blood
  Institute
National Institutes of Health

Dr. X.J. Musacchia
Graduate Programs and Research
University of Louisville

Dr. L. Muscatine, President
Department of Biology
University of California
  at Los Angeles

Peter C. Myers, Deputy Secretary
  for Agriculture
U.S. Department of Agriculture

William Nelligan
American College of Cardiology

Charles M. Oman, Ph.D., Associate
  Director
Man Vehicle Lab
Massachusetts Institute of Technology

Tobias Owen
Department of Earth and Space
  Sciences
State University of New York
  at Stony Brook

Lester Packer
Lawrence Berkeley Laboratory
University of California

Gene R. Petersen
Jet Propulsion Laboratory
California Institute of Technology

Billy J. Pfoff, Ph.D., President
Aerospace Physiologist Society

Barbara G. Pickard, Ph.D.
Biology Department
Washington University

Richard L. Popp, M.D.
Professor of Medicine
Cardiology Division
Stanford University

Dr. Frank Press
National Research Council

Dr. D.H. Reid
Aerospace Physiologist Society

Danny A. Riley, Ph.D.
Department of Anatomy
Medical College of Wisconsin

W. Eugene Roberts, D.D.S., Ph.D.
School of Dentistry
University of the Pacific

Gary F. Rockwell, M.D.
Baystate Medical Center
Wesson Women's Unit

Dr. Colin C. Rorrie, Jr., Executive
  Director
American College of Emergency
  Physicians

Carl Sagan, Ph.D.
Center for Radiophysics and Space
  Research
Cornell University

Frank B. Salisbury, Ph.D.
Plant Science Department
Utah State University

Walter Schimmerling, Ph.D.
Lawrence Berkeley Laboratory
University of California

Dr. A.L. Schuerger
Walt Disney World Co.

Alan Schwartz
Laboratory of Exobiology
University of Nijmegen, Netherlands

M. Roy Schwarz, M.D.
American Medical Association

Richard B. Searles, Secretary
International Phycological Society
Department of Botany
Duke University

John H. Siegel, M.D.
MIEMSS Shock Trauma Center
University of Maryland

Warren K. Sinclair, Ph.D.
National Council on Radiation,
  Protection, and Measurement

Thomas L. Smith, Ph.D.
Department of Physiology
Bowman Gray School of Medicine

Appendix

Gerald Sonnenfeld, Ph.D.
Department of Microbiology and Immunology
Health Sciences Center
University of Louisville

C.A. Stadd
U.S. Department of Transportation

Thomas P. Stein, Ph.D.
School of Osteopathic Medicine
University of Medicine and Dentistry of New Jersey

Paul D. Stolley, M.D., Executive Officer
International Epidemiological Association
School of Medicine
University of Pennsylvania

Scott N. Swisher, Co-Chairman
NAS Major Directions for Space Science/Life Sciences Task Group

Jill Tarter
SETI Institute

Theodore W. Tibbitts
Department of Horticulture
University of Wisconsin

Marc E. Tischler
Department of Biochemistry
University of Arizona, Health Science Center

Charles Wallach, Executive Director
International Bio-Environmental Foundation

John B. West, M.D., Ph.D.
Department of Medicine and Physiology
University of California, La Jolla

Michael L. Wiederhold
Division of Otorhinolaryngology
University of Texas, Health Science Center

M.W. Woody
Research Foundation
Ohio State University

Thomas J. Wronski, Ph.D.
Department of Physiological Sciences
University of Florida

Richard J. Wurtman, M.D.
Department of Nutrition and Food Sciences
Massachusetts Institute of Technology

Glossary

Apollo: A NASA project consisting of 17 manned flights for Earth orbital, circumlunar, and lunar missions.

Artificial gravity: Space-based simulation of the normal terrestrial gravitational field by creating a vector acceleration of 9.8 m/sec^2. See variable-gravity centrifuge.

Biocosmos: A series of Cosmos-class satellites launched by the U.S.S.R. The experiments, contributed by Soviet and international cooperators, are designed to study the effects of space flight on living organisms.

Commercially Developed Space Facility (CDSF): A spacecraft being developed by commercial partners as a permanently deployed, crew-tended space platform for materials research and manufacturing, scientific research, and storage, and as a test platform and laboratory. The craft consists of a facility module, auxiliary module, and a docking system. Astronauts will work within the pressurized "shirt sleeve" environment of the facility module during servicing; the craft will operate as an autonomous free-flier between visits.

Extravehicular activity (EVA): Activities by crew members conducted outside the pressurized hull of a spacecraft.

Free-flier: Any payload detached from another spacecraft during the operational phase of that payload and capable of independent operation.

Gemini: A NASA project consisting of 10 manned flights during 1965-1966. The project tested technologies for long-duration flight, rendezvous, docking, target vehicle propulsion, extravehicular activity, and guided reentry.

Gravity: The acceleration field associated with the mass of the Earth; approximately 9.8 m/sec^2 on the Earth's surface.

Health Maintenance Facility (HMF): A structure developed to house preventive, diagnostic, and therapeutic medical instrumentation for use on the Space Station.

Lifesat: A proposed NASA free-flier program to establish a flight and recovery capability for gravitational biology and related research.

Mainbelt asteroid: A mixture of primitive and evolved objects found in a transition region between the inner (rocky) planets and the outer (gaseous and icy) planets. The objects in this zone have apparently preserved an ordered structure related to the original temperature/pressure regime of the solar nebula.

Medicine Policy Board: A panel led by the Director of Life Sciences at NASA Headquarters and responsible for medical policies relative to the development, publication, implementation, and revision of medical standards for NASA space crews.

Microwave Observing Project (MOP): A major focus of the Search for Extraterrestrial Intelligence (SETI) Program which, when fully implemented, will permit a search for signals of natural and artificial origin over the entire sky at frequencies between 1 and 10 GHz, with a maximum sensitivity of E-10^{-23}W/m^2, and selected searches in the 1 to 3 GHz range with a maximum sensitivity of E-10^{-27}W/m^2.

Mir: Russian for "peace"; a six-port space station launched by the U.S.S.R. in 1986.

National Aerospace Plane (NASP): A joint Department of Defense/NASA program to develop and demonstrate the technologies required by a vehicle powered by airbreathing engines that would have the capability to take off and land horizontally on standard runways, cruise in the upper atmosphere at hypersonic speed, and fly directly into low-Earth orbit.

Orbital Maneuvering Vehicle (OMV): A spacecraft launched from the Shuttle Orbiter or Space Station to deploy or return free-flying payloads in low-Earth orbit.

Orbital Transfer Vehicle (OTV): An Orbital Maneuvering Vehicle capable of moving payloads between low-Earth orbit and some other orbit, typically geostationary.

Skylab: A NASA mission to study the effects of increasingly long-duration space flight, solar activity, and Earth resources. The Skylab Workshop was launched on May 14, 1973, and was visited by three Apollo astronaut crews who lived and worked in the facility for periods of 28, 59, and 84 days.

Soyuz: Russian for "union"; a spacecraft consisting of three modules: a reentry or landing module, an orbital compartment (used for crew habitation and experimentation in orbit), and an instrument compartment. Three *Soyuz* versions (original, "T," and "TM") have flown over 50 missions.

Spacehab: Commercially designed pressurized cylinders planned for incorporation into the Shuttle Orbiter payload bay and connecting to the crew compartment through the Orbiter airlock. The modules are intended for use as Orbiter middeck augmentation volumes, expanded habitation volumes, and middeck-type locker experiment facilities.

Spacelab: A general purpose, orbiting laboratory developed through the European Space Agency for crew-tended and automated activities aboard the Shuttle Orbiter. It includes both module and pallet sections, which can be used separately or in several combinations on the Orbiter.

Space Station: A platform in permanent Earth orbit for crew habitation and experimentation currently planned by NASA and international partners. Phase 1, sometimes called Block I, designates the operational Space Station, consisting of a habitation module and three laboratory modules, one for NASA, another for the European Space Agency, and the other for the National Space Development Agency of Japan. Phase 2, also known as Block II, refers to the enlarged Space Station, which will incorporate a dual keel truss structure and a servicing facility, plus additional electrical power provided by solar furnace collectors.

Variable-gravity centrifuge: A device used on orbital laboratories in which centrifugal acceleration simulates terrestrial acceleration due to gravity.

Viking: A NASA effort of 1975-1982 that consisted of two missions to Mars, each involving an orbiter and lander module. Both modules collected images of the planet; the landers also conducted chemical analyses of the Martian surface.

Selected Bibliography

National Aeronautics and Space Administration

National Aeronautics and Space Administration. Ames Research Center. April 1987. *A Profile of Life Sciences Research.* Moffett Field: Ames Research Center.

National Aeronautics and Space Administration. Ames Research Center. H. Clayton Foushee, John K. Lauber, Michael M. Baetge, and Dorothea B. Acomb. August 1986. *Crew Factors in Flight Operations: III. The Operational Significance of Exposure to Short-Haul Air Transport Operations.* NASA Technical Memorandum 88322. Moffett Field: Ames Research Center.

National Aeronautics and Space Administration. Ames Research Center. Mary M. Connors, Albert A. Harrison, and Faren R. Akins. 1985. *Living Aloft: Human Requirements for Extended Spaceflight.* NASA SP-483. Washington, DC: National Aeronautics and Space Administration.

National Aeronautics and Space Administration. Ames Research Center. Jack W. Stuster. September 1986. *Space Station Habitability Recommendations Based on a Systematic Comparative Analysis of Analogous Conditions.* NASA Contractor Report 3943. Washington, DC: National Aeronautics and Space Administration.

National Aeronautics and Space Administration. B.J. Bluth and Martha Helppie. August 1986. *Soviet Space Stations as Analogs. 2nd Edition.* NASA Grant NAGW-659. Washington, DC: National Aeronautics and Space Administration.

National Aeronautics and Space Administration. Biomedical Laboratories Branch. October 1986. *Space Station Infectious Disease Risks. Conference Report.* JSC-32104. Houston: Lyndon B. Johnson Space Center.

National Aeronautics and Space Administration. Jet Propulsion Laboratory. Thomas M. Jordan. No date given. *Radiation Protection for Manned Space Activities.* Pasadena, California: Jet Propulsion Laboratory.

National Aeronautics and Space Administration. John F. Kennedy Space Center. Biomedical Operations and Research Office. 1986. *CELSS "Breadboard" Facility Project Plan.* John F. Kennedy Space Center, FL.

National Aeronautics and Space Administration. Life Sciences Division. A.E. Nicogossian. No date given. *The USSR Space Program: Progress in Space Biology and Medicine.* Washington, DC: National Aeronautics and Space Administration.

National Aeronautics and Space Administration. Life Sciences Division. 1984. *Life Sciences Flight Program: Guide to the Life Sciences Flight Experiments Program.* No city of publication or publisher given.

National Aeronautics and Space Administration. Life Sciences Division. Space Medicine and Biological Research Branches. Donald L. DeVincenzi and Paul C. Rambaut, Study Chairpersons. September 1984. *Life Sciences: A Strategy for the 80's.* Washington, DC: National Aeronautics and Space Administration.

National Aeronautics and Space Administration. Life Sciences Division. September 1986. *Historical Overview of NASA Life Sciences Programs and Related Advisory Committee Activities.* McLean, VA: Science Applications International Corporation.

National Aeronautics and Space Administration. Life Sciences Division. December 1986. *Life Sciences Accomplishments.* Washington, DC: National Aeronautics and Space Administration.

National Aeronautics and Space Administration. Life Sciences Division. Life Sciences Strategic Planning Study Committee. Frederick C. Robbins, Committee Chairperson. 1986. *Minutes: Meeting of NASA Life Sciences Strategic Planning Study Committee, September 24-25, 1986.* McLean, VA: Science Applications International Corporation.

National Aeronautics and Space Administration. Life Sciences Division. Life Sciences Strategic Planning Study Committee. Frederick C. Robbins, Committee Chairperson. 1987. *Minutes: Meeting of NASA Life Sciences Strategic Planning Study Committee, January 22-23, 1987.* McLean, VA: Science Applications International Corporation.

National Aeronautics and Space Administration. Life Sciences Division. Life Sciences Strategic Planning Study Committee. Frederick C. Robbins, Committee Chairperson. 1987. *Minutes: Meeting of NASA Life Sciences Strategic Planning Study Committee, April 29-30, 1987.* McLean, VA: Science Applications International Corporation.

National Aeronautics and Space Administration. Life Sciences Division. Life Sciences Strategic Planning Study Committee. Frederick C. Robbins, Committee Chairperson. 1987. *Minutes: Meeting of NASA Life Sciences Strategic Planning Study Committee, August 17-18, 1987.* McLean, VA: Science Applications International Corporation.

National Aeronautics and Space Administration. Life Sciences Division. Life Sciences Strategic Planning Study Committee. Frederick C. Robbins, Committee Chairperson. 1987. *Minutes: Meeting of NASA Life Sciences Strategic Planning Study Committee, November 20-21, 1987.* McLean, VA: Science Applications International Corporation.

National Aeronautics and Space Administration. Life Sciences Division. Life Sciences Strategic Planning Study Committee. Frederick C. Robbins, Committee Chairperson. 1988. *Minutes: Meeting of NASA Life Sciences Strategic Planning Study Committee, March 11, 1988.* McLean, VA: Science Applications International Corporation.

National Aeronautics and Space Administration. Los Alamos National Laboratories. September 1986. *Manned Mars Missions. A Working Group Report. Revision A.* Ed. Michael B. Duke and Paul W. Keaton. NASA M001. No city of publication given: National Aeronautics and Space Administration.

National Aeronautics and Space Administration. Lyndon B. Johnson Space Center. Space and Life Sciences Directorate. 1987. *An Overview of Life Sciences Programs, Organization, and Resources in the Space and Life Sciences Directorate, Johnson Space Center.* Selected materials presented to NASA Life Sciences personnel at NASA Headquarters on October 28, 1986. Washington, DC: Science Applications International Corporation.

National Aeronautics and Space Administration. Lyndon B. Johnson Space Center. Scientific and Technical Information Office. Michele Anderson. *BIOSPEX: Biological Space Experiments. A Compendium of Life Sciences Experiments Carried on U.S. Spacecraft.* NASA Technical Memorandum 58217. Houston: Lyndon B. Johnson Space Center. June 1979.

National Aeronautics and Space Administration. Lyndon B. Johnson Space Center. 1985. *Life Sciences Flight Experiments Program: Life Sciences Laboratory Equipment (LSLE) Descriptions.* Houston: Lyndon B. Johnson Space Center.

National Aeronautics and Space Administration. Lyndon B. Johnson Space Center. Space Station Projects Office. July 1986. *Medical Requirements of an Inflight Medical System for Space Station.* Houston: Lyndon B. Johnson Space Center.

National Aeronautics and Space Administration. Lyndon B. Johnson Space Center. Michele Anderson, Pauline Posey, and John Lintott. 1986. *Life Sciences Experiments in Space: JSC Life Sciences Project Division.* Houston: Lyndon B. Johnson Space Center.

National Aeronautics and Space Administration. No date given. *NASA Educational Opportunities in the Life Sciences. Space: The New Frontier.* No city of publication or publisher given.

National Aeronautics and Space Administration. November 1978. *Guidelines for Acquisition of Investigations.* NHB 8030.6A. Washington, DC: National Aeronautics and Space Administration.

National Aeronautics and Space Administration. Office of External Relations. Technology Utilization and Industry Affairs Division. James J. Haggerty. July 1984. *Spinoff 1984.* Washington, DC: U.S. Government Printing Office.

National Aeronautics and Space Administration. Office of Commercial Programs. James J. Haggerty. August 1985. *Spinoff 1985.* Washington, DC: U.S. Government Printing Office.

National Aeronautics and Space Administration. Office of Commercial Programs. James J. Haggerty. August 1986. *Spinoff 1986.* Washington, DC: U.S. Government Printing Office.

National Aeronautics and Space Administration. Office of Space Science and Applications. Life Sciences Division. *Biological and Medical Experiments on the Space Shuttle 1981-1985.* Ed. Thora W. Halstead and Patricia A. Dufour. Washington, DC: National Aeronautics and Space Administration.

National Aeronautics and Space Administration. Office of Space Science and Applications. 1987. *Space Life Sciences Symposium: Three Decades of Life Science Research in Space. Abstracts. Washington, D.C. June 21-26, 1987.* No city of publication given: National Aeronautics and Space Administration.

National Aeronautics and Space Administration. Office of Space Science and Applications. Life Sciences Division. A.E. Nicogossian. October 1984. *Human Capabilities in Space.* NASA Technical Memorandum 87360. Washington, DC: National Aeronautics and Space Administration.

National Aeronautics and Space Administration. Office of Space Science and Applications. Life Sciences Division. October 1985. *CELSS (Controlled Ecological Life Support Systems) Program: Research Projects Funded 1979-1985.* Washington, DC: National Aeronautics and Space Administration.

National Aeronautics and Space Administration. Sally K. Ride. August 1987. *Leadership and America's Future in Space; A Report to the Administration.* Washington, DC: National Aeronautics and Space Administration.

National Aeronautics and Space Administration. Scientific and Technical Information Branch. Office of Space Science and Applications. Life Sciences Space Station Planning Committee. 1986. *Life Sciences Space Station Planning Document: A Reference Payload for the Life Sciences Research Facility.* NASA Technical Memorandum 89188. Washington, DC: National Aeronautics and Space Administration.

National Aeronautics and Space Administration. Scientific and Technical Information Branch. John A. Pitts. 1985. *The Human Factor: Biomedicine in the Manned Space Program to 1980.* The NASA History Series. NASA SP-4213. Washington, DC: U.S. Government Printing Office.

National Aeronautics and Space Administration. Scientific and Technical Information Branch. Office of Space Science and Applications. Global Biology Science Working Group. Mitchell B. Rambler, Working Group Chairperson. 1983. *Global Biology Research Program: Program Plan.* Ed. M.B. Rambler. NASA Technical Memorandum 85629. Washington, DC: National Aeronautics and Space Administration.

National Aeronautics and Space Administration. Scientific and Technical Information Branch. Office of Space Science and Applications. Life Sciences Division. 1985. *Life Sciences Accomplishments.* NASA Technical Memorandum 88177. Washington, DC: National Aeronautics and Space Administration.

National Aeronautics and Space Administration. Scientific and Technical Information Branch. 1985. *Search for the Universal Ancestors.* Ed. H. Hartman, J.G. Lawless, and P. Morrison. NASA SP-477. Washington, DC: U.S. Government Printing Office.

National Aeronautics and Space Administration. Scientific and Technical Information Branch. SETI Science Working Group. John H. Wolfe and Samuel Gulkis, Working Group Chairpersons. 1983. *SETI Science Working Group Report.* Ed. Frank Drake, John H. Wolfe, and Charles L. Seeger. NASA Technical Paper 2244. Washington, DC: National Aeronautics and Space Administration.

National Aeronautics and Space Administration. Scientific and Technical Information Branch. Study Group on the Cosmic History of the Biogenic Elements and Compounds. John A. Wood, Study Group Chairperson. 1985. *The

Cosmic History of the Biogenic Elements and Compounds. Ed. John A. Wood and Sherwood Chang. NASA SP-476. Washington, DC: U.S. Government Printing Office.

National Aeronautics and Space Administration. Scientific and Technical Information Branch. Science Workshops on the Evolution of Complex and Higher Organisms. David Raup, Workshops Chairperson. 1985. *The Evolution of Complex and Higher Organisms.* Ed. David Milne, David Raup, John Billingham, Karl Niklaus, and Kevin Padian. NASA SP-478. Washington, DC: U.S. Government Printing Office.

National Aeronautics and Space Administration. Scientific and Technical Information Branch. Donald P. Hearth, Study Director. January 1976. *Outlook for Space: Report to the NASA Administrator by the Outlook for Space Study Group.* NASA SP-386. Washington, DC: National Aeronautics and Space Administration.

National Aeronautics and Space Administration. Scientific and Technical Information Branch. Patricia A. Dufour, Judy L. Solberg, and Janice S. Wallace. 1985. *Publications of the NASA CELSS (Controlled Ecological Life Support Systems) Program.* NASA Contractor Report 3911. Washington, DC: National Aeronautics and Space Administration.

National Aeronautics and Space Administration. Scientific and Technical Information Branch. Janice S. Wallace and Donald L. DeVincenzi. 1986. *Publications of the Exobiology Program for 1984: A Special Bibliography.* NASA Technical Memorandum 88382. Washington, DC: National Aeronautics and Space Administration.

National Aeronautics and Space Administration. SESAC Task Force on Scientific Uses of Space Station. Peter M. Banks, Task Force Chairman. 1986. *Space Station Summer Study Report.* Ed. David C. Black and Hugh S. Hudson. No city of publication or publisher given.

National Aeronautics and Space Administration. Space Biomedical Research Institute. Lyndon B. Johnson Space Center. May 1987. *Results of the Life Sciences DSOs Conducted Aboard the Space Shuttle 1981-1986.* Ed. Michael U. Bungo, Tandi M. Bagian, Mark A. Bowman, and Barry M. Levitan. Houston: Lyndon B. Johnson Space Center.

National Aeronautics and Space Administration Advisory Council

National Aeronautics and Space Administration Advisory Council. Earth System Sciences Committee. Francis P. Bretherton, Committee Chairperson. January 1988. *Earth System Science: A Closer View.* Washington, DC: National Aeronautics and Space Administration.

National Aeronautics and Space Administration Advisory Council. Life Sciences Advisory Committee. November 1978. *Future Directions for the Life Sciences in NASA.* Ed. G. Donald Whedon, John M. Hayes, Harry C. Holloway, John

Spizizen, and S.P. Vinograd. Washington, DC: National Aeronautics and Space Administration.

National Aeronautics and Space Administration Advisory Council. Solar System Exploration Committee. Committee Chairpersons: David Morrison (1983-86), Noel Hinners (1981-82), and John Naugle (1980-81). 1983. *Planetary Exploration Through Year 2000: A Core Program. Part 1.* Washington, DC: U.S. Government Printing Office.

National Aeronautics and Space Administration Advisory Council. Solar System Exploration Committee. Committee Chairpersons: David Morrison (1983-86), Noel Hinners (1981-82), and John Naugle (1980-81). 1986. *Planetary Exploration Through Year 2000: An Augmented Program.* Ed. Beven M. French. Washington, DC: U.S. Government Printing Office.

National Aeronautics and Space Administration Advisory Council. Solar System Exploration Committee. Committee Chairpersons: David Morrison (1983-86), Noel Hinners (1981-82), and John Naugle (1980-81). 1986. *Planetary Exploration Through Year 2000: A Core Program: Mission Operations.* Washington, DC: U.S. Government Printing Office.

National Aeronautics and Space Administration Advisory Council. Space and Earth Science Advisory Committee. Louis J. Lanzerotti, Committee Chairperson. November 1986. *The Crisis in Space and Earth Science: A Time for a New Commitment.* No city of publication or publisher given.

National Academy of Sciences

National Academy of Sciences. National Research Council. Committee on Earth Sciences. R.G. Prinn, Committee Chairperson. 1985. *A Strategy for Earth Science from Space in the 1980's and 1990's. Part II: Atmosphere and Interactions with the Solid Earth, Oceans, and Biota.* Washington, DC: National Academy Press.

National Academy of Sciences. National Research Council. Committee on Planetary Biology and Chemical Evolution. Lynn Margulis, Chairperson. 1981. *Origin and Evolution of Life — Implications for the Planets: A Scientific Strategy for the 1980's.* Washington, DC: National Academy of Sciences.

National Academy of Sciences. National Research Council. Committee on Space Biology and Medicine. Jay M. Goldberg, Committee Chairperson. 1987. *A Strategy for Space Biology and Medical Science for the 1980s and 1990s.* Washington, DC: National Academy Press.

National Academy of Sciences. National Research Council. Committee on the Space Station. Robert C. Seamans, Committee Chairperson. September 1987. *Report of the Committee on the Space Station of the National Research Council.* Washington, DC: National Academy Press.

National Academy of Sciences. National Research Council. Space Science Board. Assembly of Mathematical and Physical Sciences. Neal S. Bricker, Study Chairperson. 1979. *Life Beyond the Earth's Environment: The Biology of Living Organisms in Space.* Washington, DC: National Academy of Sciences.

National Academy of Sciences. National Research Council. Space Science Board. Committee on Planetary Biology. Daniel D. Botkin, Committee Chairperson. 1986. *Remote Sensing of the Biosphere.* Washington, DC: National Academy Press.

National Academy of Sciences. National Research Council. Space Science Board. Donald B. Lindsley, Study Chairperson. 1972. *Human Factors in Long-Duration Spaceflight.* Washington, DC: National Academy of Sciences.

National Academy of Sciences. National Research Council. Space Science Board. Kenneth V. Thimann, Study Chairperson. 1970. *Space Biology.* Washington, DC: National Academy of Sciences.

National Academy of Sciences. National Research Council. Space Science Board. 1970. *Life Sciences in Space.* Ed. H. Bentley Glass. Washington, DC: National Academy of Sciences.

National Academy of Sciences. National Research Council. U.S. Committee for an International Geosphere-Biosphere Program. John S. Eddy, Committee Chairperson. Commission on Physical Sciences, Mathematics, and Resources. Herbert Friedman, Commission Chairperson. 1986. *Global Change in the Geosphere-Biosphere: Initial Priorities for an IGBP.* Washington, DC: National Academy Press.

Federation of American Societies for Experimental Biology

Federation of American Societies for Experimental Biology. July 1983. *Research Opportunities in Cardiovascular Deconditioning: Final Report Phase I.* Ed. M.N. Levy and J.M. Talbot. NASA Contractor Report 3707. Washington, DC: National Aeronautics and Space Administration.

Federation of American Societies for Experimental Biology. July 1983. *Research Opportunities in Space Motion Sickness: Final Report Phase II.* Ed. J.M. Talbot. NASA Contractor Report 3708. Washington, DC: National Aeronautics and Space Administration.

Federation of American Societies for Experimental Biology. April 1984. *Final Report Phase III. Research Opportunities in Bone Demineralization.* Ed. S.A. Anderson and S.H. Cohn. NASA Contractor Report 3795. Washington, DC: National Aeronautics and Space Administration.

Federation of American Societies for Experimental Biology. April 1984. *Final Report Phase IV. Research Opportunities in Muscle Atrophy.* Ed. G.J. Herbison and J.M. Talbot. NASA Contractor Report 3796. Washington, DC: National Aeronautics and Space Administration.

Federation of American Societies for Experimental Biology. December 1985. *Research Opportunities on Immunocompetence in Space.* Ed. William R. Beisel and John M. Talbot. NASA Contractor Report 176482. Washington, DC: National Aeronautics and Space Administration.

Federation of American Societies for Experimental Biology. January 1985. *Research Opportunities in Human Behavior and Performance.* Ed. Julien M. Christensen and John M. Talbot. NASA Contractor Report 3924. Bethesda, MD: Federation of American Societies for Experimental Biology.

Federation of American Societies for Experimental Biology. Life Sciences Research Office. 1986. *Research Opportunities in Nutrition and Metabolism in Space.* Ed. Philip L. Altman and Kenneth D. Fisher. Contract Number NASW 3924. Bethesda, MD: Federation of American Societies for Experimental Biology.

U.S. Congress

U.S. Congress. House of Representatives. Committee on Science and Technology. Don Fuqua, Committee Chairperson. October 1983. *National Aeronautics and Space Act of 1958, as Amended, and Related Legislation.* 98th Congress. 1st Session. Committee Print.

U.S. Congress. Office of Technology Assessment. 1985. *International Cooperation and Competition in Civilian Space Activities.* Washington, DC: U.S. Government Printing Office.

U.S. Congress. Office of Technology Assessment. Thomas F. Rogers, Project Director. November 1984. *Civilian Space Stations and the U.S. Future in Space.* OTA-STlO241. Washington, DC: U.S. Government Printing Office.

Other Published Reports

Conklin, James J., and Richard I. Walker, eds. 1987. *Military Radiobiology.* Orlando, FL: Academic Press.

National Commission on Space. Thomas O. Paine, Committee Chairperson. May 1986. *Pioneering the Space Frontier.* New York: Bantam Books.

Pitts, John A. 1985. *The Human Factor: Biomedicine in the Manned Space Program to 1980.* NASA SP-4213. Washington, DC: National Aeronautics and Space Administration.

University of Texas Medical Branch. 1983. *Life Sciences Experiments for a Space Station: A Compilation of the Reports of Seven Working Groups.* Ed. Jill D. Fabricant. Galveston, TX: University of Texas Medical Branch.

Periodicals

Bungo, Michael W., John B. Charles, and Philip C. Johnson, Jr. 1985. Cardiovascular Deconditioning During Space Flight and the Use of Saline as a Countermeasure to Orthostatic Intolerance. *Aviation, Space, and Environmental Medicine* 56 (October):985-990.

Dixon, G.A., J.D. Adams, and W.T. Harvey. 1986. Decompression Sickness and Intravenous Bubble Formation Using a 7.8 Psia Simulated Pressure-Suit Environment. *Aviation, Space, and Environmental Medicine* 57 (March):223-228.

Hills, B.A. 1985. Compatible Atmospheres for a Space Suit, Space Station, and Shuttle Based on Physiological Principles. *Aviation, Space, and Environmental Medicine* 56 (November):1052-1058.

Ilyn, E.A. 1983. Investigations on Biosatellites of the Cosmos Series. *Aviation, Space, and Environmental Medicine* 12, sec. 2 (December):S9-Sl5.

Ivanov, Boris and Olga Zubareva. 1985. To Mars and Back Again on Board Bios. *Soviet Life* (April):22-25.

Johnson, P.C. 1979. Fluid Volumes Changes Induced by Spaceflight. *Acta Astronautica* 6:1335-1341.

Klein, Harold P. 1981. U.S. Biological Experiments in Space. *Acta Astronautica* 8:927-928.

LeBlanc, A., C. Marsh, H. Evans, P. Johnson, V. Schneider, and S. Jhingran. 1985. Bone and Muscle Atrophy with Suspension of the Rat. *Journal of Applied Physiology* 58:1669-1675.

Parker, D.E., M.F. Reschke, A.P. Arrott, J.L. Homick, and B.K. Lichtenberg. 1985. Otolith Tilt-Translation Reinterpretation Following Prolonged Weightlessness: Implications for Preflight Training. *Aviation, Space, and Environmental Medicine* 56 (June):601-606.

Stassinopoulos, E.G. 1980. The Geostationary Radiation Environment. *Journal of Spacecraft and Rockets* 17 (March-April):145-152.

Tucker, C.J., I.Y. Fung, C.D. Keeling, and R.H. Gammon. 1986. Relationship Between Atmospheric CO_2 Variations and a Satellite-Derived Vegetation Index. *Nature* 319 (January):1-5.

Additional Materials

NASA and Department of Defense (DOD). March 9, 1983. *Memorandum of Agreement Between NASA and DOD on Life Sciences Activities in Support of the Space Transportation System (STS).*

NASA and United States Air Force (USAF). September 21, 1983. *Memorandum of Agreement Between NASA and USAF on Life Sciences Activities in Support of Space Transportation System (STS).*

SB/Life Sciences Division. January 22, 1981. NASA Management Instruction 8900.1A. Subject: *Operational Medical Responsibilities for the Space Transportation System (STS).*

Appendix

Photograph Credits

Page Number	Description	Reference
8	Vestibular Sled	NASA LBJ S81-39883
10	Man-on-the-Moon	NASA 75-HC-I59
41	Blood Draw	NASA LBJ S09-05-0143
54	Solar Flare	NASA 72-HC-749
61	Solar Particle Event	NASA 85-HC-148
68	WETF	NASA LBJ S86-35712
73	Exercise in Space	NASA LBJ S09-07-0520
80	Manipulator Arm	NASA 85-HC-484
83	JSC Spacesuit	NASA S86-36740
83	ARC Spacesuit	NASA AC87-0662-284
92	Health Maintenance Facility	NASA LBJ S87-29982
102	Centrifuge	NASA AC87-0281-44
104	Pine Seedlings	NASA 85-HC-441
115	Breadboard Facility	NASA KSC-387C-525/2
120	Production Chamber	NASA KSC-387C-1060/5
125	Ozone Hole	NASA 86-HC-289
126	Guayaquil, Ecuador	NASA 84-HC-534
128	Mississippi Delta	Eosat Company Photo
133	Martian Surface	NASA 80-H-605
146	Saturn Montage	NASA 83-HC-221
155	*Challenger*	NASA 85-HC-167
172	Space Station Mockup	NASA MSC 780331
189	Agriculture from Space	Eosat Company Photo

Index

Agriculture, Department of, 127, 130, 171, 177, 182
Air Force, U.S., 171, 177, 182
    School of Aerospace Medicine, 40, 55
    Office of Scientific Research, 40
    Space Command, 51
American Heart Association, 51
American Institute of Biological Sciences, 175-76
Announcement of Opportunity, 36, 48, 159, 176, 197
Antarctic, 67, 74, 89, 92, 125, 149
Apollo, 8, 174, 201
    crew, 12, 41, 45-46, 202
    program, 17, 27
artificial gravity (see variable-gravity centrifuge),
astronaut medical information data base, 7, 13, 30-31, 94, 98-99

Biocosmos, 109, 154, 158-60, 201
Biosatellite, 4, 14, 27, 158
Breadboard Project, 20, 113-17, 119-21, 212
Brookhaven National Laboratory, 54

Challenger, 10, 14, 21, 155, 157-58, 164, 173-74, 212
Columbia University, 54
Comet Halley, 135, 138, 147
Comet Rendezvous Asteroid Flyby Mission, 135-36, 147
Commerce, Department of, 91
Commercial Centers for the Development of Space, 15, 185-87, 190
Commercially Developed Space Facility, 109, 154, 158-59, 201
Congress, U.S., 173, 178
countermeasures, to the effects of microgravity on humans, 6, 18, 27, 32, 42, 44, 46, 48-49, 97, 99, 107, 161-62
Crew Emergency Return Vehicle, 7, 13, 19, 29, 31, 93, 97-98

deconditioning, as a result of extended exposure to microgravity
    physiological, 5-6, 18, 29, 41-43
    psychological, 5, 18, 72-73
Defense, Department of, 65, 91, 156-57, 177, 182, 202
Detailed Supplementary Objective, 95, 176
Discipline Working Groups, 36, 159, 181
Dryden Flight Research Facility, National Aeronautics and Space Administration, 91

Earth Observing System, 9, 33-34, 126, 128, 130
Energy, Department of, 127, 130, 171, 177, 182
England, 39
Environmental Protection Agency, 127, 130
Europa, 146
European Space Agency, 39, 50, 119, 130, 160, 202
expendable launch vehicle, 36, 108, 158, 168
extravehicular activity, 4, 7-8, 12, 27, 29-30, 73, 79, 81-84, 88, 96, 99, 177, 201

Federal Aviation Administration, 91
Federal Communications Commission, 91
Federal Coordinating Committee in Science and Technology, 50
Federal Laboratory Consortium, 185-86
Federal Republic of Germany (West Germany), 36, 39, 158
Federal Technology Transfer Act of 1986, 185-86
France, 39
French National Center for Space Studies, 160

galactic cosmic rays (see radiation)
Gemini, 12, 46, 201
German Research and Development Institute for Air and Space Travel, 119, 160

Health Maintenance Facility, 7, 13, 19, 29, 31, 86, 92-94, 97-98, 160, 188-90, 201, 212

India, 149
Intergovernmental Personnel Act, 183-84
International Geosphere-Biosphere Programme, 14, 124, 127-29

Japan, 5, 36, 119, 122, 149
Jet Propulsion Laboratory, National Aeronautics and Space Administration, 54, 196
Jupiter, 146

Lawrence Berkeley Laboratory, 54, 62
Lawrence Livermore Laboratory, 54
Lifesat, 154, 201
Life Sciences Division (see NASA Divisions)
Life Sciences Strategic Planning Study Committee (see NASA Advisory Council)
life support requirements, during extended space flight, 7, 30, 88, 112
life support systems, 2, 4, 7, 20, 27, 30, 82, 88, 112, 182

Mars, ix, 1, 6-9, 12-18, 20-21, 28, 31, 33-35, 40, 44, 50, 59-60, 67, 69, 81, 86-88, 92-93, 95, 97-98, 102, 105, 107, 120, 122, 133-34, 137-38, 140-47, 153-54, 161, 163, 166, 197, 202, 212
Mars Observer Mission, 137
Mars Rover/Sample Return Mission, 137, 147
Medicine Policy Board, National Aeronautics and Space Administration, 31, 91, 201
microgravity, 7, 31-32, 40, 42-47, 49, 67-68, 86, 89, 94, 97, 101-04, 106-07, 109, 116, 119-21, 135-36, 147, 162, 188, 197
Microwave Observing Project (*also see* Search for Extraterrestrial Intelligence), 4, 9, 27, 35, 144, 148-49, 201
Mir, 109, 202
Moon, ix, 1, 8, 13-16, 35, 58, 60, 81-82, 86-87, 93, 102, 107, 137, 140, 145, 166

Appendix

National Academy of Sciences, 34–35, 127, 179
    Astronomy Survey Committee, 150
    Committee on Planetary Biology and Chemical
        Evolution, 34, 134
    Committee on Space Biology and Medicine, 91, 101
National Aeronautics and Space Act of 1958, 154, 185
NASA Advisory Council, ix, 1, 5, 34, 127
    Earth System Sciences Committee, ix, 124
    Life Sciences Advisory Committee, 91, 126
    Life Sciences Strategic Planning Study Committee, ix,
        1–2, 4–5, 11, 14, 22–25, 27, 34, 39, 128, 179, 191–96
    Solar System Exploration Committee, ix, 149
NASA Centers, 11, 15, 38, 40, 47, 92, 112, 118, 121, 171–74,
    180–81, 183–84, 187, 190
    Ames Research Center, 39–40, 79, 88, 92, 101–02, 112–14,
        116–17, 121, 171, 195, 212
    Goddard Space Flight Center, 29, 54
    Johnson Space Center, 10, 39–40, 54, 79, 88, 91–95, 97,
        101, 112, 118, 121, 171, 176, 195, 212
    Kennedy Space Center, 40, 91–92, 94, 101, 112–16, 121
    Langley Space Center, 54
    Marshall Space Flight Center, 79, 102
NASA divisions, 166
    Astrophysics Division, 9, 14, 34, 137, 150, 171
    Communications Division, 171
    Earth Science and Applications Division, 9, 14, 33–34,
        125, 127–28, 149, 171
    Life Sciences Division, ix, 3, 9, 11, 13–17, 22, 26, 31, 34,
        37–38, 47–48, 79, 91, 109–10, 122, 124–25, 128, 132, 137,
        140, 144–45, 148–50, 152, 154, 156, 158, 165, 171–84, 187,
        189–90
    Microgravity Science and Applications Division, 171
    Solar System Exploration Division, 14, 33–34, 137, 140,
        144–45, 150, 171
NASA Headquarters, 11, 15, 38–40, 47, 54, 79, 88, 91–92, 101,
    112, 171, 174–75, 180–82, 184, 187, 190, 195, 201
NASA initiatives, under consideration
    Exploration of the Solar System, 9, 17, 34
    Humans to Mars, 17
    Mission to Planet Earth, 9, 17, 20, 34, 124
    Outpost on the Moon, 17
NASA Long Range Planning Committee for the Life
    Sciences, 47
NASA Offices
    Office of Aeronautics and Space Technology, 79, 88,
        171, 178, 182
    Office of Commercial Programs, 171, 186–87, 190
    Office of Equal Opportunity Programs, 171
    Office of Management, 171
    Office of Space Flight, 79, 91, 171, 182
    Office of Space Science and Applications, 5, 33, 88, 91,
        132, 137, 140, 144–45, 148–50, 171, 173, 176, 178
    Office of Space Station, 79, 171, 182, 187–90
National Aerospace Plane, 92–93, 98, 202
National Commission on Space, 17, 102
National Council of Radiation Protection, 55

National Institutes of Health, 37, 40, 48, 50–51, 65, 156, 167,
    171, 177, 179, 182–83
National Library of Medicine, 11, 38
National Oceanic and Atmospheric Administration, 14, 55,
    65, 126–27, 130, 156
National Science Foundation, 14, 48, 127, 130, 149–50, 156,
    165, 171, 177, 182
    Division of Polar Programs, 149
National Space Development Agency of Japan, 39, 50, 119,
    130, 160, 202
National Space Policy, 2, 17
Naval Medical Research and Development Command, 40
Naval Research Laboratory, 55

Oak Ridge National Laboratory, 55
Office of Management and Budget, 173, 178
Orbital Maneuvering Vehicle, 93, 202
Orbital Transfer Vehicle, 93, 202

Phillips, General Samuel E., 174
Program Operating Plan, 172, 174
psychological factors, during extended space flight, 12,
    67–76, 85, 95, 154

radiation, 53–65, 87, 103, 116, 121, 161–63, 168, 191, 197
    biological effects of ionizing radiation, 5, 28, 55, 62–65,
        158
    galactic cosmic rays, 6, 12, 18, 53, 55–56, 58–60, 62, 64
    HZE (high atomic number, Z, and high energy, E), 18,
        58, 60–62, 64
    linear energy transfer, 12, 55–56, 60–64
    relative biological effectiveness, 55, 63–64
    shielding from, 6, 56, 58–62, 64–65, 82
    solar particle events, 6, 12, 18, 53–56, 58–63, 212
Reagan, President Ronald, 2, 17, 186
Request for Proposals, 48, 176, 197
Research and Technology Operating Plan, 172, 174
Ride, Sally K., 17, 102, 124
Romanenko, Yuri, 6, 18
Rosetta/Comet Nucleus Sample Return Mission, 135–36,
    147

Salyut, 109
Saturn, 17, 138, 147, 212
Science Applications International Corporation, 196
Scientific and Technical Information Facility, National
    Aeronautics and Space Administration, 38, 187, 190
Search for Extraterrestrial Intelligence (*also see* Microwave
    Observing Project), 27, 35, 133, 144, 148–50, 197, 201
Skylab
    crews, 12, 41, 45–47
    missions, 56, 202
solar particle events (*see* radiation)
Soviet Space Agency, 50
Soviet Union (U.S.S.R.), 5–6, 36, 39, 105, 109, 118–19, 122,
    149, 158, 160–61, 179, 183, 201, 202

Soyuz, 12, 46, 109, 202
space adaptation syndrome (also known as space motion sickness), 6, 12, 41–44, 49, 197
spacecraft maximum allowable concentrations, 96
Spacehab, 109, 154, 158–59, 202
Space Infra Red Telescope Facility, 136
Spacelab, 3, 10, 14, 21, 26, 36, 41, 61, 73, 104, 108–09, 154–55, 158–59, 163, 168, 173, 202
space motion sickness (*see* space adaptation syndrome)
Space Pallet Satellite, 154, 158
Space Shuttle, 4, 10, 14, 21–22, 27, 36, 41–42, 61, 80, 93, 96, 98, 105, 109, 122, 129, 154, 158–60, 162, 164, 168, 202
Space Station, 4–5, 7, 17, 25, 27–30, 35–36, 40, 50, 56, 67, 70, 74, 81–89, 92–94, 97–99, 102, 105, 108–09, 119–20, 122, 129, 135–36, 147, 154, 156–57, 159–63, 166–68, 176, 182, 188, 201–02, 212
    Phase 1, 3, 10, 26, 36
    Post-Phase 1, 3, 26
Space Station Cosmic Dust Collection Facility, 137, 147
Space Station Gas Grain Simulation Facility, 137, 147
spacesuit, 83, 88, 96, 99, 212

Space Transportation System, 31, 92, 108, 157–58
Specialized Center of Research, 11, 14, 38, 50, 180, 183
Stevenson-Wylder Technology Innovation Act of 1980, 185
Sun, 54, 130, 145

Titan, 17, 35, 137–38, 140, 147
Titan/Cassini Mission, 137–38, 147
Transportation, Department of, 91

Universities Space Research Association, 95
University of California, 55
University of Maryland, 94
University of San Francisco, 55

variable-gravity centrifuge, 8, 12–13, 19, 29–30, 32, 46, 49, 86, 97, 101–02, 107–08, 163, 168, 178, 182, 201–02, 212
Viking, 21, 137, 146, 202
Voyager, 138, 146

White House Office of Science and Technology Policy, 131, 177, 182

Acknowledgements

The NASA Advisory Council wishes to thank the many contributors to this publication:

- Frederick C. Robbins, M.D., Chairperson of the Life Sciences Strategic Planning Study Committee (LSSPSC)
- Members of the LSSPSC
- Staff Associates of the LSSPSC
- James H. Bredt, Ph.D., Executive Secretary of the LSSPSC
- Maurice M. Averner, Ph.D., Alternate Executive Secretary of the LSSPSC
- J. Richard Keefe, Ph.D., Scientific Advisor to the LSSPSC
- Domenic A. Maio, Ph.D., Science Applications International Corporation (SAIC) Project Manager
- Joyce A. Douglass, SAIC Deputy Project Manager
- Abby A. Johnson, Ph.D., Technical Writer and Editor
- Kathleen A. Dunk, Production Coordinator
- Margaret I. Siriano, Researcher
- Peter G. Gustafson, Graphics Manager
- Paul Jenkins, Artist

Abbreviations and Acronyms

AIBS	American Institute of Biological Sciences
AO	Announcement of Opportunity
ARC	Ames Research Center
CDSF	Commercially Developed Space Facility
CELSS	Controlled Ecological Life Support Systems
CERV	Crew Emergency Return Vehicle
CNES	Centre National d'Etudes Spatiales (National Center for Space Studies-French)
CRAF	Comet Rendezvous and Asteroid Flyby
DFRF	Dryden Flight Research Facility
DFVLR	Deutsche Forschungs- and Versuchsanstalt fuer Luft- und Raumfahrt (German Research and Development Institute for Air and Space Travel)
DOD	Department of Defense
DOE	Department of Energy
DSO	Detailed Supplementary Objective
E(r)	Earth radius
ELV	Expendable launch vehicle
EOS	Earth Observing System
EPA	Environmental Protection Agency
ESA	European Space Agency
EUE	Experiment Unique Hardware
EVA	Extravehicular activity
g	Gravity
GCR	Galactic cosmic rays
GEO	Geosynchronous Earth orbit
HMF	Health Maintenance Facility
HZE	High atomic number, Z, and high energy, E
IGBP	International Geosphere-Biosphere Programme
IOC	Initial Operating Configuration
IPA	Intergovernmental Personnel Act
ISM	Interstellar medium
JSC	Johnson Space Center
KSC	Kennedy Space Center
LEO	Low-Earth orbit
LET	Linear energy transfer
LSLE	Life Sciences Laboratory Equipment
LSSPSC	Life Sciences Strategic Planning Study Committee
MOP	Microwave Observing Project
MSIS	*Man-Systems Integration Standards*
NAC	NASA Advisory Council
NAS	National Academy of Sciences
NASDA	National Space Development Agency of Japan
NASP	National Aerospace Plane
NIH	National Institutes of Health
NOAA	National Oceanic and Atmospheric Administration
NSF	National Science Foundation
OAST	Office of Aeronautics and Space Technology
OCP	Office of Commercial Programs
OMB	Office of Management and Budget
OMV	Orbital Maneuvering Vehicle
OSF	Office of Space Flight
OSSA	Office of Space Science and Applications
OTV	Orbital Transfer Vehicle
PMC	Permanently Manned Capability
POP	Program Operating Plan
psi	Pounds per square inch
R&T	Research and Technology
RBE	Relative biological effectiveness
REM	Roentgen equivalent in man
RFP	Request for Proposals
RNA	Ribonucleic acid
RTOP	Research and Technology Operating Plan
SCOR	Specialized Center of Research
SETI	Search for Extraterrestrial Intelligence
SIRTF	Space Infrared Telescope Facility
SMAC	Spacecraft maximum allowable concentrations
SPAS	Space Pallet Satellite
SPE	Solar particle event
STS	Space Transportation System
USAF	U.S. Air Force
USDA	U.S. Department of Agriculture
USRA	Universities Space Research Association

www.ingramcontent.com/pod-product-compliance
Lightning Source LLC
Chambersburg PA
CBHW081722170526
45167CB00009B/3668